OFF THE BEATEN TRACK IN ICELAND

Jón G. Snæland

OFF THE BEATEN TRACK IN ICELAND

SKRUDDA
2012

Maps made by the National Land Survey of Iceland (NLSI) in this book are derived from the institutes' IS 50V database with the exception of paths covered in this book that come from its author. The data is presented as it is in the NLSI database at the time the book was made. **As a result of the Icelandic government's focus to prevent illegal off-road driving and to restrict or prevent traffic in certain areas of the country, NLSI wants to take notice of that the users of the published data are responsible for monitoring if and when roads and trails are closed temporarily or permanently to public traffic. Such government measures reach further than information on roads and tracks that are in the NLSI databases at any time.** Thus, the NLSI shall not be liable for direct or indirect loss or damage as a result of the use of data from the NLSI or of derived data, except to the extent that would fall under the Product Liability Law no. 25/1991. In other respects, the NLSI declines liability to the extent permitted in law.

Off the beaten track in Iceland

Layout and design: Skrudda
English translation: Michael J. Kissane
Maps: National Land Survey of Iceland
Type: Adobe Garamond Pro 10,5/13,5
Printed in Slovenia
Published by Skrudda
skrudda@skrudda.is

ISBN: 978–9979–655–87-9

Contents

Introduction

How to drive in Iceland

Driving in Iceland requires closer attention and more expertise than do most other places. There is much to be wary of, even in the lowlands and built up areas. Paved roads have a way of disappearing suddenly to become gravel roads which are often so narrow as to require slowing down, even stopping on the hard shoulder, when meeting another vehicle. There is also the danger of accident or damage due to small rocks shooting out from under the wheels of fast-moving cars. Gravel roads are often literally covered with loose gravel which tends to get pushed to the centre and sides of the road. When driven into, these ridges can have the same effect as snow and ice: the vehicle slips and slides and all too often goes off the road. There are also many "blind hills" on gravel roads that require a slow approach, especially since the crest of the hill may well be hiding a sharp turn just beyond.

The highlands of Iceland can be dangerous if drivers fail to get full information in advance about the roads and current driving conditions or if they don't show due care. The most-travelled highland roads are marked with an F on most maps (e.g. F26 – Sprengisandsvegur). The F indicates that the Icelandic Public Roads Administration sees to keeping up the road in question. However, the road or track may simply be opened and scraped in early summer, no more, and thus its condition can deteriorate suddenly. There are hardly any warning signs on the highland roads, and signposts are fairly rare, so travellers really must have a good road map with them. Though the F-roads are in the care of the Public Roads Administration most have few if any bridges, so the numerous watercourses must be forded. Highland roads in Iceland are often closed well into June, even July, because of snow and frost-melt quagmires on the roads. The Icelandic Public Roads Administration's web-site includes a map showing which roads are open or closed and the condition of the roads in the road system. *www.vegagerdin.is/english, http://www.vegagerdin.is/vegakerfid/Fjallvegir/.*

There are a great number of tracks in the highlands and other wilderness areas that are not under the Public Roads Administration. These are unmarked tracks with no road signs and often not maintained at all. Commonly they are simply two tyre tracks and can be impassable for vehicles other than super-jeeps driven by highly experience mountain men and even then Icelanders only travel on such tracks in groups, with a number of jeeps.

Fording rivers

Travellers try various tricks when they reach a ford, such as throwing many stones into the river to assess its depth. The trouble with this method is that it may not show up any deeper channels in the river, or quicksand or boulders. On the whole, this is an unsafe method, though it may help in assessing the current. The only real way to find out whether a ford is safe is to wade across, testing the bottom for hidden deeper channels. Things can go very wrong if a vehicle gets stuck for any length of time in a river with a fast current, even if it isn‘t deep. This is because the current can undermine the wheels incredibly quickly, sinking the vehicle even deeper. Until recently rivers were simply considered impassable if they could not be waded across, but that was while jeeps had smaller wheels. Today it is not uncommon to drive across unwadable rivers because of how much larger wheels are now, 44 inch tyres are, for instance, 110 cm high. On the other hand, this can lead to another danger: In rivers with a heavy current, driving straight athwart the current can bring such pressure on these large wheels that the jeep starts drifting. This means the passengers are in great danger if the river is unwadable.

In the "Arctic Truck Jeep Book" there is a section on tyres floating. There it says that a vehicle loses 1 kg of weight for each volume liter that goes in the water. Thus it becomes lighter by about one ton if one cubic meter goes under water. Example: A Land Cruiser 120 on 38" tyres weighs about 2.5 tons including passengers and luggage. 38" tyres have a volume of 200 litres each and each housing could be 30 litres. This means that four tyres and housings under water lighten the car in water by almost 900 kg. When the car body goes under it lightens very quickly and the current starts to really make a difference. Example: When the floor of a Land Cruiser 120 goes under the car lightens by 450 kg. for the first 15 cm. If it then goes even deeper the area of the rear is added which means it lightens by 600 kg. for each 10 cm. A car like that which is in water half way up the doors then weighs only about 500 kg. (Source: Jeppabók Arctic Trucks).

Always take good care when fording a river. Photo: Jón G. Snæland.

When driving into a river it is important to have selected the right gear. You really do not want to change gears in deep water, since there is a danger of water getting in and causing the clutch to slip. The best thing, therefore, is to get into low gear before entering the water.

Off-road driving harms the land. Photo: Þóra Sigurbjörnsdóttir.

Always try to drive a little in a downstream direction if at all possible, and at the same speed as the current.

Glacial rivers require even more care. If the river appears daunting it is often better to drive across early in the morning before the sun-melt has started to affect its volume, probably between 05:00 and 08:00.

Some good advice

1. Turn back if you aren't completely sure you'll get across.
2. If you are in the role of leader, it often calls for more courage to turn away from a river than to try to drive across. Turn back rather than endanger the lives of your fellow travellers.

From the Conservation of Nature Laws from 1999, no. 44, March 22.

Off-road driving

It is forbidden to drive machine-driven vehicles off road. However, such vehicles may be driven on glaciers, also on snow outside of built-up areas on condition that the ground is frozen and snow-covered.

3. Do not try to increase your speed when you can no longer hear the engine.
4. Do not look into the water for any length of time.
5. Do not depend entirely on the people on the river bank.
6. Fasten a tow-rope to the jeep before entering the river.
7. Get the car waterproofed, be aware of the air intake and how high it is.

From the Icelandic Road Administration; *ust.is*

Off-road driving seems to be a growing problem in Iceland. Iceland‘s many-faceted nature is one of the main attractions for those of us who travel round the country. There are very many sensitive places so careless driving can cause wounds in the countryside that can take years, even decades, to recover. In our travels here we need to be guided by respect for the country‘s nature. The soil is volcanic which means it is so crumbly that the wheels of even one vehicle can easily create deep tracks whether the soil is grassy or bare. It is extremely difficult to get rid of damage caused by off-road driving. Iceland being just south of the Arctic Circle, where vegetation has a very short growing time, any damage to vegetation can take many decades to recover. The same is true of sand and bare ground, where scars can take even longer to disappear since there is no vegetation to eventually cover it. Tyre tracks ruin the look of the land, besides which they become channels for water and thus contribute to erosion. Off-road tracks also attract other drivers and encourage even more destructive driving.

Stick to driving on roads and marked tracks. Turn around or walk on where driving is no longer viable. Be informed about the route you plan to take. Use road maps to organise your trip before you set out.

Respect seasonal restrictions to traffic in the highlands and other wilderness areas and keep track of bulletins about the state of roads. Keeping a responsible attitude to nature means everyone will have the chance to enjoy it for years to come.

Off-road driving is totally forbidden on pain of fines and imprisonment!

From the Icelandic Road Administration

Most mountain roads and roads in the interior of Iceland have a surface of loose gravel. The same applies to some sections of the national highway, which also has long stretches of asphalt. The surface on the gravel roads is often loose, especially along the sides of the roads, so one should drive carefully and slow down whenever approaching an oncoming car. The mountain roads are also often very narrow, and are not made for speeding. The same goes for many bridges, which are only wide enough for one car at a time. In addition to not having an asphalt surface, the mountain roads are often very winding. Journeys therefore often take longer than might be expected. For information on road conditions, Tel.: +354-1777, daily 8:00-16:00 The total length of the Ring Road around Iceland (national highway nr. 1) is 1.339 km.

The general speed limit is 50 km/h in urban areas, 80 km/h on gravel roads in rural areas, and 90 km/h on asphalt roads.

Please note:

Special warning signs indicate danger ahead, such as sharp bends, but there is generally not a separate sign to reduce speed. Please choose a safe speed according to conditions. Motorists are obliged by law to use headlights at all times day and night. In Iceland all driving off roads or marked tracks is prohibited by law. Passengers in the front and backseats of an automobile are required by law to use safety-belts. Icelandic law forbids any driving under the influence of alcohol.

Motor vehicle insurance:

A “Green Card” or other proof of third-party insurance is mandatory for motorists driving their own cars in Iceland, except from the following countries: Austria, Belgium, Croatia, Channel Islands, the Czech Republic, Cyprus, Denmark, Faroe Islands, Finland, France, Germany, Gibraltar, Great Britain, Greece, Hungary, Ireland, Isle of Man, Italy, Liechtenstein, Luxembourg, Monaco, Netherlands, Northern Ireland, Norway, Portugal, San Marino, Slovakia, Slovenia, Spain, Sweden, Switzerland and the Vatican.

Filling stations:

In the greater Reykjavík area most filling stations are open every day to 23:30. Opening hours around the country, where the pumps are privately operated, can vary from place to place. Many stations in the Reykjavík area and larger towns of Iceland have automats in operation after closing, which accept visa and euro credits cards as well as notes.

Opening of mountain tracks:

Most mountain roads are closed until the end of June, or even longer because of wet and muddy conditions which make them totally impassable. When these roads are opened for traffic many of them can only be negotiated by four-wheel-drive vehicles. For some mountain tracks it is strongly advised that two or more cars travel together. Also, before embarking on any journey into the interior collect as much information as possible regarding road conditions from a travel bureau, tourist information office or the Public Roads Administration, tel.: +354-1777 daily 8:00-16:00

Emergency Telephone (Fire, Police, Ambulance) 112
http://www.112.is

Hornbjarg
Ísafjarðardjúp
BOLUNGARVÍK
SUÐUREYRI
HNÍFSDALUR
Drangajökull
FLATEYRI
ÍSAFJÖRÐUR
SÚÐAVÍK
HÚNAFLÓI
SKAGAFJÖRÐUR
SIGLUFJÖRÐU
ÓLAFSFJÖ
Arnarfjörður
ÞINGEYRI
GLÁMA
SKAGASTRÖND
HOFSÓS
ÁRSKÓ
BÍLDUDALUR
PATREKSFJÖRÐUR
HÓLMAVÍK
SAUÐÁRK
BLÖNDUÓS
VARMA
Flatey
HVAMMSTANGI
LAUGARBAKKI
Blanda
BREIÐAFJÖRÐUR
STYKKISHÓLMUR
BÚÐARDALUR
HELLISSANDUR
RIF
GRUNDARFJÖRÐUR
ÓLAFSVÍK
Snæfellsjökull
Arnarvatnsheiði
Norðurá
Eiríksjökull
LANGJÖKULL
Kjölur
BORGARNES
FAXAFLÓI
AKRANES
Gullfoss
Geysir
REYKJAVÍK
SELTJARNARNES
LAUGARVATN
Þingvallavatn
GARÐUR
ÁLFTANES
KÓPAVOGUR
FLÚÐIR
SANDGERÐI
KEFLAVÍK
GARÐABÆR
LAUGARÁS
NJARÐVÍK
HAFNARFJÖRÐUR
HVERAGERÐI
VOGAR
Hekla
HAFNIR
SELFOSS
EYRARBAKKI
GRINDAVÍK
ÞORLÁKSHÖFN
HELLA
STOKKSEYRI
HVOLSVÖLLUR
VESTMANNAEYJAR
Surtsey
Dyrhólaey
162
70
102
13
93
121
17
82
97
21
67
139
25
118
45
90
29
48
78
36
167
900 m
600 m
300 m
100 m
0
0 10 20 30 40 50 60 70 80 90 100 km

The black and red squares show the approximate area of each route. The numbers refer to the appropriate page numbers.

On the route to Arnarvatnsheiði. Mt. Strútur to the left and Eiríksjökull glacier to the right. Photo: Steingrímur Steinþórsson.

Arnarvatnsheiði (Skagfirðingavegur)

Start – Finish	Skeljaháls F578 – Kjalvegur, F35.
Distance/Time	85.1 km. Day trip.
Fuel	At Húsafell.
Caution needed due to	Fording and large rocks.
Passable for	All full-sized jeeps.
Of interest	Caves, rivers, lava fields, fishing in lakes and rivers.

This route starts at the farm Kalmanstunga at Skeljaháls pass. Drive along the mountain Lambafell and initially follow the lava of Kleppahraun, but later the track enters Hallmundarhraun. Hallmundarhraun has many caves, among them Víðgemlir, Stefánshellir, Hallmundarhellir, Kalmanshellir and Surtshellir. Surtshellir is the best-known of these and is right by the road. It is 1500 metres long and has four openings down to the cave. Ancient bones have been found in this cave. Legends say that outlaws lived in the caves in the 10th century, but that they were killed by residents of Borgarfjörður. After driving some distance past Surtshellir, the road splits, with the left track leading over Norðlingafljót river and inland to Úlfsvatn lake. The ford at Norðlingafljót can be somewhat deep along the south

Crossing the river Norðlingafljót by Helluvað ford. These travellers are returning from up on the heath. Photo: Jón G. Snæland.

bank. The track inland to Úlfsvatn lake is fairly passable as far as the fishermen's hut, as much effort has recently been spent on improvement. It is possible to drive onwards, past Úlfsvatn lake and arrive at Arnarvatnsvegur F578, north of Jónsvatn lake, but this track has large rocks and is slow, tedious going in unmodified jeeps.

The usual route lies straight ahead and on to the Helluvað ford on the Norðlingafljót river. It is possible to drive this route with all standard vehicles. Helluvað ford can have a lot of water in it, but the bottom is solid, though rocky.

An old stone-built farm building at Álftakrókur. Núpavatn lake to the right and beyond it Mt. Strútur. Photo: Jón G. Snæland.

Ólafur's cairns, or The Bishop's cairns were built by the men of Ólafur Hjaltason, bishop at Hólar, one autumn in 1552-69. They got caught in a blizzard and each man was made to pile up a cairn to keep from freezing to death. Photo: Jón G. Snæland.

North of the ford is a longish stretch of entrenched dirt road which can be tough going during wet weather. There is hut for the shepherds of Borgarfjörður at Álftakrókur, and it is possible to book space here for the night. Two tracks lead from here up the heath. One is called Leggjarbrjótur (Leg-breaker), and is a very rough route, today mostly used by horsemen. The other one, the main option, lies from the hut along the Mordísarhæð heights and inland to Réttarvatn lake. There is a road branch leading further in to Arnarvatn stóra (Larger Arnarvatn) lake, and onward down to Miðfjörður fjord. This route is quite passable. On the north side of Larger Arnarvatn lake is the Hnúabaksskáli hut, where accommodation and fishing permits are sold.

Skagfirðingavegur continues between the lakes of Réttarvatn and Larger Arnarvatn, and across a fairly large stream called Skammá which runs from Réttarvatn. Just a bit further inland, the Búðará river is forded. Beyond the ford, the track becomes somewhat slow-going, despite occasional sections of good road. The route here is newly posted by the touring group Ferðaklúbburinn 4x4 all the way east to Kjalvegur. These posts are a big help in finding the right way through the area. To the north of Skagfirðingaleið is a marked track called Grímstunguheiðarvegur which goes to the Fellaskáli hut, and further northward in to the Víðidalur valley. Under no conditions is it advisable to take this route, as it is essentially impassable to all vehicles, due to mud and runoff. Farther east one arrives at a cluster of cairns called Ólafsvörður (Ólafur's cairns) or Biskupsvörður (the Bishop's cairns) at Stórisandur. These were erected by the men of Ólafur Hjaltason, bishop at Hólar in north Iceland (1552–1569), when they were caught in bad weather one autumn and tried to keep warm through the effort of building cairns. Each man built his own cairn, and the road goes between the cairns. Once past the cairns, the road becomes quite slow-going and rocky until one approaches Grettishæð height, where there is a junction with the road of Grímstunguheiði heath and the road along the Krákshraun lava field. Grettishæð height is marked on maps, and said to be 765 metres above sea level. It is here that a man known as Þorbjörn öngull is said to have buried the head of Grettir, a hulk of a Viking and an outlaw. From here the route goes along Krákshraun lava eastward to Kjalvegur. Parts of this stretch are often unclear and slow-going, and there is not much likelihood of meeting travellers on this part.

Ásgarðsleið

Start – Finish	**Kjalvegur, F35, inland from the Innri-Skúti – Kerlingarfjallaskálar huts.**
Distance/Time	**10 km. 30-40 minutes' drive.**
Fuel	**Kerlingarfjöll during summers and at Geysir during winters.**
Caution needed due to	**Fording of the Blákvísl branch and streams.**
Passable for	**All vehicles during summers.**
Of interest	**Colourful mountains, waterfalls and hot springs.**

From Kjalvegur, F35, turn into the Ásgarðsleið track at a sign that is marked "Kerlingarfjöll 10 km". Soon after the turn, one crosses a creek called Fossrófulækur, which is no obstacle, and then continues up a slope until a flat area with an airfield is reached. Once past the airfield, one goes down a slope to the ford of the Blákvísl branch, where it joins another branch, Jökulkvísl (Jökulfallið), which Sveinn Pálsson (1762–1840), a doctor and biologist, over-simplified to Fallið in his diaries from the years 1791–1797. The Blákvísl ford is quite rough on the bottom, so drive slowly when crossing the stream.

The Kerlingarfjöll mountains bathed in the setting sun. Photo: Bjarni Kristinsson.

Just below the ford is a pretty waterfall, Gýgjarfoss, in the Jökulkvísl river branch which runs from Hofsjökull glacier. The Jökulkvísl branch here drops into a chasm and there are more falls downstream in the chasm, which can be seen by going down the cleft beneath the previously mentioned airfield.

The old Ferðafélag Íslands lodge in the Kerlingarfjöll mountains. Photo: Jón G. Snæland.

A borehole, used as a natural bathing pool, near the Ásgarður hut. Photo: Bjarni Kristinsson.

Above Gýgjarfoss waterfall, follow Jökulkvísl branch up slopes until finally arriving at a bridge over the branch, built in 1936. There is a waterfall here called Hvinur. Upon completion of the bridge, the route was considered passable for vehicles going in to the Kerlingarfjöll mountains. But before the bridge was built, the branch was forded somewhat upstream from where the bridge is now, and the route went along a track still clearly visible. Across the bridge is an intersection where a left turn is made in to the Kerlingarfjöll mountains, whereas going straight leads down into Leppistungur and a new track down to Skipholtskrókur.

Skipholtskrókur was in the news in 2007 when a search was made for the "Holy Grail," which some, including the Italian Giancarlo Gianazza, believed had been hidden by the Knights of the Temple in Skipholtskrókur. The Knights of the Temple was an order of monks, named for the Temple of King Solomon in Jerusalem. The Knight's mission was, among other things, to protect various Christian treasures, such as the Holy Grail from which Jesus drank during the Last Supper.

From the intersection, drive up a slope where there is another airfield, and then drive down to the Ásgarðsá river. Here it is possible to cross on a large culvert that has been set in the river. From here, drive up a rather steep hill, which then reveals a beautiful valley and a cluster of huts at the base of Mt. Ásgarðsfjall. There is a rough track to the left, which goes along Illahraun lava field and onward emerging onto the Fjórðungssandur sands and down along the river Þjórsá, on the Sprengisandsvegur, F26, at the Sultartangavirkjun hydro-electric power plant.

The tourism company Fannborg operates various services in the Kerlingarfjöll mountains. In addition to a variety of overnight accommodations, there is a bath-house with showers and three hot tubs, and sale of fuel. The campground has sites for tents and camper trailers, with access to running water and toilets. There is also a licensed restaurant in the main lodge. Usual open season is from June 15th to September 30th. The buildings here are lighted and heated by electricity from Sælufossvirkjun power plant on the Ásgarðsá river about 400 metres from the buildings. A shepherds' hut was in the Kerlingarfjöll mountains from 1890 and it stood at the eave of the mountain hut built by the touring club Ferðafélag Íslands in 1936. In 1961, people began considering the possibility of enjoying skiing in the area, and then built a ski school.

Driving up the valley and repeatedly crossing the Ásgarðsá river, we arrive at a bore hole, which supplies warm water for a small pool made at the site. It is possible to drive up beyond the huts in the direction of the mountain called Snækollur and get close to an alpine glacier that tops this mountain.

Búðarháls

Start – Finish	**Bridge on the Tungnaá River – The waterfall called Dynkur.**
Distance/Time	**28 km. Two hours' drive.**
Fuel	**At Hálendismiðstöð at Hrauneyjar.**
Caution needed due to	**None.**
Passable for	**All jeeps.**
Of interest	**Waterfalls, canyon and great views.**

Turn onto the route across a rather new bridge over the Tungnaá river, at a point just above the old auto and livestock cable-ferry over to Búðarháls. The bridge was built in conjunction with the Búðarhálsvirkjun hydro-electric power plant. Drive up a ridge and then turn north along a high ridge on a good and rather clear track. Having reached an altitude of 664 metres on a small hillock, inland from Hálsamót hill, there is a good view to either side – across the Tungnaá river on the one hand, and across the Þjórsá river on the other – as well as views to six different glaciers.

The tracks on Búðarháls are poorly if at all marked, and therefore it is almost a certainty that those not following a map carefully or lacking a GPS navigation device will become somewhat lost here. The old route along Búðarháls lay entirely right above the river, and was graded shortly after the cable-ferry over the Tungnaá river was built in 1964, with the road from the site called Hald inland to the ford at Tryppavað on the Tungnaá coming about 1966.

Driving across the Kaldakvísl ford east of Búðarháls. Photo: Jón G. Snæland.

The route goes downhill again, passing a track to the right, and coming to another intersection, where a turn is made to the left towards the Þjórsá river and Gljúfurleitarfoss waterfall. This route is somewhat tough going and reaches nearly to the edge of the Þjórsá canyon, where it is a short walk to see the waterfall. From the edge of the canyon, return the same way, taking the first track to the left to drive to the waterfall Dynkur (Búðarhálsfoss). The track does not reach all the way to the waterfall, and one must walk the last part to the waterfall.

It is also possible to drive back up the ridge, along the track and down the hill to an intersection. Take the left, in the direction of the Þjórsá river, and we then arrive at the Þjórsá above the waterfall Dynkur. From here it is also a short distance to the waterfall. When this route is driven, we can see the shepherds' hut Hvanngiljahöll, which is two kilometres north of the road. From this hut it is possible to either drive or walk another five kilometres upstream to a third waterfall, called Hvanngiljafoss (Kjálkaversfoss).

Before the time of the cable-ferry and the bridge built in 1964, very few travelled in the Búðarháls area, since it is between two large glacier rivers. Such little travel there and so much isolation from other sections of the country led to differing names for the same waterfalls, east and west

From the Þjórsá gorge above the waterfall Dynkur. Photo: Jón G. Snæland.

of the rivers. To the west, the residents of Gnúpverja district (and most Icelanders) call one of the area's large waterfalls Dynkur, but on the east, members of the Holt district call the same waterfall Búðarhálsfoss. In the

The waterfall Dynkur (Búðarhálsfoss). Photo: Stefanía Guðjónsdóttir.

old days, the only way across Búðarháls was by ferry boat at the old ferry site called Hald, just below the bridge. This was where sheep were ferried across, and even an auto in 1933, when the first auto was driven across the Sprengisandur sands.

Djúpavatnsleið

Start – Finish	**Route 42 north of Kleifarvatn lake – On Ísólfsskálavegur 427**
Distance-Time	**25 km. A little over an hour.**
Fuel	**In the town of Hafnarfjörður.**
Caution needed due to	**No particular dangers on the way.**
Passable for	**All vehicles in summer.**
Of interest	**Beautiful lava fields, lakes and many old eruption craters.**

Djúpavatnsleið lies right across the Reykjanes Preserve, which was put on the protection list in 1975. The park is about 300 km^2 and is the largest preserve in Iceland.

From the town of Hafnarfjörður, drive briefly southwest towards the airport on Route 41 to the intersection with Route 42 which we take left, towards Kleifarvatn lake. When we drive at the foot of Háuhnúkar, we come to a gravel quarry and are headed towards the Vatnsskarð gap. As the road starts up the slope, across from the quarry there is an intersection with a road we take to the right (F428), a short ways towards the moutain Fjallið eina, where there are many popular hiking trails. This is the actual start of Djúpavatnsleið.

Djúpavatnsleið, as the route is called, runs southwesterly between the mountain Sandfell to the west and mountain Sandfellsklofi to the east.

A big landslide on the south slopes of Núpshlíðarháls ridge. Photo: Jón G. Snæland.

We drive at the base of a slope known as Sveifluháls, where a large rockslide occurred in the earthquake of 2000, and continue further south on the Hofmannaflöt flats, where we drive right across lava and arrive at a mountain east of Djúpavatn lake. It is also possible to drive straight along the base of the slope to Ketilsstígur, which then links to Djúpavatnsleið north of the Vigdísarvellir plain.

Northeast of Djúpavatn is a track leading off to the right from Djúpavatnsleið, which goes to Lækjarvellir plains. On Lækjarvellir in the valley of Móhálsdalur there is a recently built and pleasant hikers' hut, which the Association of Community Governments in the Capital Region funded. From the hut it is a short distance to Djúpavatn lake. At the south end of the lake is the anglers' cabin for members of the Anglers' Association of Hafnarfjörður. A fair amount of Arctic charr is caught in the lake.

Djúpavatn is a crater lake, 0.15 square kilometres in size at an elevation of 195 metres above sea level. Despite the name, which means 'Deep Lake' it is only 16.7 metres deep in the deepest part.

At Vigdísarvellir, our next destination, there are ruins of an abandoned farm, at the foot of the rise Núpshlíðarháls (Vesturháls). The farm was built about 1830 and was lived in until 1905, when the house collapsed during an earthquake. South of Vigdísarvellir is a row of fissure craters, where we can seen the route winding southward. The road was made for the laying of a fibre optic cable and has now been closed.

Southernmost on Djúpavatnsleið at the base of the ridge Núpshlíðarháls, we drive along a pretty valley bordered by a row of cliffs formed by a cata-

The hut at Djúpavatn. It is open to the public. Photo: Jón G. Snæland.

strophic natural event. Exiting the valley, it is a short way to Ísólfsskálaleið, where our journey ends.

During summers, Djúpavatnsleið is passable for standard autos, without noteworthy hindrances, except for the occasional large rock in the road or huge mud puddles, which can be numerous during wet weather.

Dyngjufjalladalur

Start – Finish	**The farm Svartárkot – F910 south of Kattbekingur, south-west of Askja.**
Distance/Time	**63 km. Five to six hours' drive.**
Fuel	**At Reykjahlíð on the shores of Lake Mývatn.**
Caution needed due to	**Rough lava and very slow-going road; easy to get lost.**
Passable for	**All modified jeeps and those with high-clearance.**
Of interest	**Beautiful rivers, vast lava fields and desert and majestic mountains.**

The start of this route is at the farm Svartárkot, which is the southernmost farm in the district of Suður-Þingeyjarsýsla, and one of the highest altitude farms in the county, at over 400 metres elevation. Drive past the farm and soon we arrive at one of the most deeply entrenched highland roads in Iceland. At this point the route is called Suðurárbotnavegur, and for quite a while it follows the Suðurá river, a large spring-fed river that emerges in full volume from the Suðurá bottoms. At many points the road is dirty and entrenched more than a metre lower than the surrounding terrain. In

Kattbekingur is a tuff mountain particularly noticeable in its surroundings, the pale brown mountain rising from the grey-black of lava and sands. At its foot there are many large boulders that have tumbled from the mountain and it is almost unnatural to see the brown boulders lying scattered on the black sand. Photo: Jón G. Snæland.

One of two footbridges over the river Suðurá. Photo: Jón G. Snæland.

dry weather, lots of fine dust covers the road, whereas in wet weather the road is quickly spoiled. In such places, it is also difficult to meet oncoming vehicles, due to tight conditions.

There are beautifully vegetated gravel bars in the Suðurá and opposite the river the Suðurárhraun lava is rough, sandy and appears intimidating. Two simple bridges were built over the Suðurá river in 1960, and were intended to allow livestock and people to cross. There are shepherds' huts near the bridge at Stóra-Flesja, which are generally available to travellers. Upstream from the huts, drive from the river towards Mt. Sellandafjall. Here we drive along lots of eroded terrain and hummocks, up against which sheep often seek good shelter.

After several kilometres of slow-going track, we arrive at a track to the right, which we take. This is the so-called Lákaleið, in honour of Þorlákur Páll Jónsson of the farm Garður in the Mývatnsveit district who posted this part of the route. The track is slow-going along the Útbrunahraun lava, all the way to the hut Botn, at the source of the Suðurá river. At this point it is ideal to take a 10 minute walk to the source of the Suðurá. From Botn, drive further through Útbrunahraun and along high road shoulders at Frambrunahraun lava. Here the track is very unclear and very tortuous,

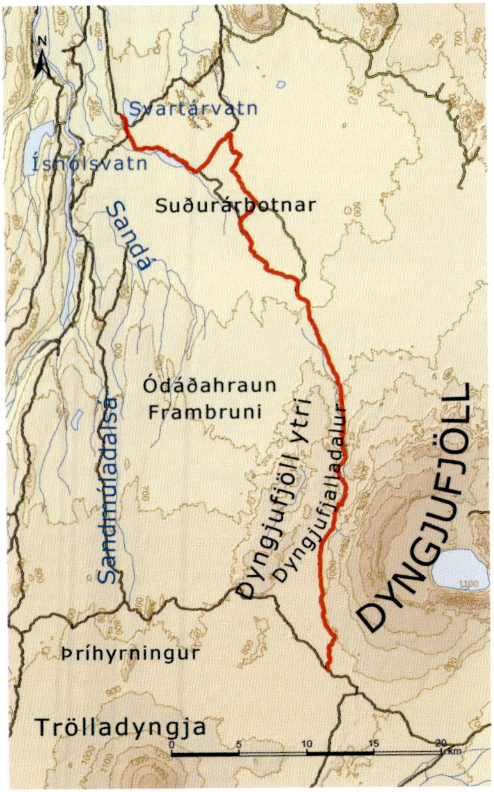

and much of the time is spent in first gear in lowest drive.

Now head north in the valley and out onto Útbrunahraun lava, which initially is fairly easy going, but once we approach the Suðurá bottoms, travel is only in first gear and low drive for several kilometres. The part under the rough margin of Frambrunahraun lava can be hazardous for smaller jeeps, and there is a risk of damage to vehicles, due to high lava bumps that stick up from the track. But gradually the track becomes better, and when we approach the valley Dyngjufjalladalur, we finally emerge onto the sands.

Dyngjufjalladalur is a narrow valley, sandy and barren. Some who have travelled through the area compare it to the "Wild West," and the palagonite cliffs there bear an undeniable resemblance. A small stream that is a blend of free-stone and glacial water runs through the valley, with growing volume in the afternoons, but which is not an obstacle during summertime. Northernmost in the valley is Dyngjufellsskáli, a hut run by the touring club Ferðafélagið Akureyrar. Several site names in the neighbourhood harken back to pioneers of earlier times, like Lokatindur (Locktind), which is named for the Englishman Lock, also called 'Reindeer Lock'. Mt. Vatnsfell, with an elevation 1,308 metres above sea level and southwest of Askja, was named after another Englishman, Watt, a lawyer and pioneer who, among other things, named a mountain in Vatnajökull glacier, Pálsfjall, on an expedition across Vatnajökull glacier. However, Vatnajökull is not named after this same man, but rather refers to vast amounts of water that

At this point the road is more than two meters lower than the surrounding terrain. This is common along this road, which is mostly soil, so in dry weather a lot of fine dust forms on it. This means driving soon becomes difficult in wet weather. It can also be problematic for vehicles to pass each other between the high banks. Photo: Jón G. Snæland.

run from it in all directions. Dyngjufjalladalur was first driven by Sigurjón Rist and Páll Arason in 1944 in a 1931 Ford. From the Dyngjufellsskáli, follow the valley upstream between the Dyngjufjöll ytri (Outer Dyngjufjöll Mountains) and Dyngjufjöll Mountains, and drive east of Fjallsendi and out onto a large flatland at Kattbekingur, where the route ends on F910, on Trölladyngjuleið (Dyngjufjallaleið).

At the source of the river Suðurá at the Suðurá bottoms, only ten minutes walk from the Botni hut. Photo: Jón G. Snæland.

Emstruleið

Start – Finish	**Fljótshlíð, Route 261 – Route 210.**
Distance/Time	**35 km. Two hours' drive.**
Fuel	**At Hvolsvöllur.**
Caution needed due to	**Glacial and free-stone rivers.**
Passable for	**All jeeps.**
Of interest	**Glaciers and canyons.**

Innermost on the valley slope Fljótshlíð, below the farm Fljótsdalur, we reach the track called Emstruleið. Here we drive onto the gravels of the Markarfljót river, and the first obstacle is the river Þórólfsá, which can be crossed by ordinary vehicles if larger ones are included in the group. Regular vehicles have little business on the Emstruleið, except possibly as far as the banks of the Gilsá river. Just upstream from Þórólfsá is the track leading past the hut called Fell, which is upslope from the road. The track continues inland at the foot of Mt. Þórólfsfell, and is soon squeezed between the mountains and the Markarfljót River. The river has often flooded this area and cut out the section of road at the foot of Mt. Þórólfsfell. In earlier years, the road was above Mt. Þórólfsfell, where the old postal route went above the abandoned farm Kanastaðir. It is possible for modified jeeps to go this route, about five kilometres upwards from Mt. Þórólfsfell, west of

A horseback rider crosses the Gilsá River. The old road is visible in the cliffs above. Photo: Jón G. Snæland.

the Tjarnarmýri moors. Inland from Streitur, the routes separate, and one track goes out onto the gravels, while the other skirts the foot of the cliffs. The route along the foot of the cliffs has not been maintained, and is therefore quite tough-going due to rockslides and the hill's slope. The route onto the gravels leads directly to the Gilsá river, at a reasonably good ford. But sometimes the river is split into several channels over the gravels. The Gilsá, which is a glacial river, has its source in Tindfjallajökull glacier, and can be an obstacle on hot or wet days, and has often proven a hindrance for winter travellers as well. Inward along the slopes there is also a ford across the Gilsá, where the current can be swift and the banks high and undercut. Across the gravels is an intersection and from there it is possible to drive gently upslope and eventually arrive at a pretty field called Hellisvellir, where there was once a shepherds' hut, which is now merely ruins. Nearby is a round-up corral for sheep, tucked into the foot of a high cliff. It is also possible to drive inland to a gap called Fauskheiði, and a good, clearly visible track lies all the way in to a point across from the site of Bólstaður beneath the prominence called Einhyrningur. Start to drive around Fausksheiði, and soon we can see across the Markarfljót river and onward to the valley Húsadalur in Þórsmörk Preserve. Sometimes thoroughly experienced jeep drivers and river men cross the river here, but the less experienced should not even attempt it.

Continue driving along Fauskheiði and then along the Tröllagjá canyon, a natural wonder, magnificent any way you look at it. Tröllagjá is thought to have been formed in a major flood about 2,500 years ago, following a

A traveller in the Markarfljót gorge just below the bridge at Emstrur. Photo: Jón G. Snæland.

volcanic eruption under the glacier Mýrdalsjökull. Next, we arrive at the flats Einhyrningsflatir, at the foot of the prominence Einhyrningur, where the sizeable shepherds' hut, called Bólstaður, serves farmers of Fljótshlíð. This is a perfect place to have your snack on the grass in front of the hut. Beneath Einhyrningur, along the creek that falls from Hrútholt above, are the ruins of an old farmstead. The centuries' old volume *Landnámabók*, (The Book of Settlement), says "... Sighvatur the Red lived at a place called Bólstaður ..."

Resuming travel, we go up a steep hill inland of Einhyrningur, and follow the mountainsides. These can have considerable snow, well into spring, and due to the slanting slopes, the hills are often impassable while snow persists there. In these cases, it may be worthwhile going eastward, where the slope is appreciably less.

Inland from the shoulders of Einhyrningur is a clearly visible track down to Markarfljót canyon and to Svartikrókur. Once over the 'shoulder' of Einhyrningur, go down to the Markarfljót river, but note that during summers there can be considerable runoff in the slopes down to the river.

To the right of the slopes is a very powerful waterfall that is worth a look. Drive over the river on the bridge built in 1978, which replaced an old cable-ferry established in 1913 and still visible just below the bridge. Before the cable-ferry, a boat was used for crossing the river at one of its calmest points, at the so-called *Eyrar* beneath the mountain Stóra-Grænafjall. Livestock, men and their horses were loaded aboard the boat, which was sometimes pulled between banks with a rope. Shortly before 1880, two men drowned here while rowing a boat during such an effort.

Upstream from the bridge over the Markarfljót (now called Gamla brúin, "Old bridge") stands a shepherds' hut, Mosar, and an older hut that is now used for livestock. Inwards of the huts, one crosses a small lava slab approaching Mt. Hattfell. North of Hattfell is an intersection, and a track there leads south of Hattfell and about five kilometres down to the hut run by the touring club Ferðafélag Íslands at Efri-Botn, and onwards to Hattfellsgil chasm.

Formerly people drove across the river Fremri-Emstruá and down to Þórsmörk, and then southward from Mt. Hattfell along sandy terrain, where the track disappears every year. Members of the group Minnsta

Barbara Ósk Ólafsdóttir drives a modified Musso 4x4 with 38-inch tyres across the Markarfljót river in a great deal of current. The airfield at Húsadalur is in the background. Photo: Þóra Sigurbjörnsdóttir.

In the book, *Ekið um óbyggðir* an accident is described as having taken place at the Emstruá bridge site. It says, "A foreign photographer had come east of the Mælifellssandur sands, to the Emstruá river, with two hitch-hikers who had gotten a ride from him shortly before. They arrived at the river, which was essentially in flood conditions. The photographer thought the bridge was only for pedestrians and livestock, and decided to try and drive across the river, but the hitch-hikers were terrified and decided to stay behind on the river bank. The river immediately took the vehicle (a Lada Sport) with it, but the photographer managed to get out of the auto and was rescued by the hitch-hikers. The Lada was carried under the bridge and over a waterfall, where it was visible for many days after the accident."

ferðafélagið made the first trip east of Hattfell and across the Fremri-Emstruá (Syðri-Emstruá) and down to Húsadalur in 1961. This route is very difficult and dangerous. They drove west of Mófellshnaus and down a very steep hill to the river. This hill has proven too much for some and there have been many incidents of overturned cars and vehicle damage.

A jeep trip made on August 18, 1979 proved fatal in the river, and trips

A small, standard version 4-wheel drive vehicle crosses the branch Bláfjallakvísl. Photo: Jón G. Snæland.

Mt. Hattfell seen from Svartikrókur. The cliffs in the foreground are the walls of Markarfljót Canyon. Photo: Jón G. Snæland.

along this route virtually ceased. From the previously named intersection near Hattfell, drive in front of the Útigönguhöfði bluff, to the river Innri-Emstruá, which was bridged in 1975.

North of the Fremri-Emstruá, another stream branch Bláfjallakvísl is forded. The ford is good, and the bottom solid, but there is always considerable volume in the branch. The route ends a short distance north of the ford at an intersection east of Mælisfellssandur and the track in to Hvanngil canyon.

Emstruleið can be difficult during winters and there can be obstacles at various places. A common problem is the Gilsá river, which can be difficult since the river is somewhat below the bank at the crossing point. The hill above Einhyrningur is also very difficult when snow fills the roadway. Before taking this option, it may prove better to drive east of the bend around the hill. Bláfjallakvísl is also often open, and with high snow banks.

Eyjafjarðarleið

Start – Finish	Hólsgerði in Eyjafjörður – Laugafell.
Distance/Time	37 kilometres. One and a half to three hours' drive.
Fuel	In the town of Akureyri.
Caution needed due to	Some steep hills, otherwise, nothing serious.
Passable for	All vehicles with high clearance, but not caravans to the south.
Of interest	Majestic mountain, beautiful valley, and wilderness.

Follow Route 821 south along Eyjafjörður as far as Hólsgerði, which is the start of the mountain road F821. Soon after the start of F821, we drive to the very start of Eyjafjörður, with the local river Eyjafjarðará cascading along falls and rapids. The road cuts often into the mountainside and many small creeks and springs either cross the road, like small rivers, or run along the roadside for goodly distances. There is a lot of water along the way, but it poses no obstacle.

Deep into the Eyjafjarðardalur valley, at the site called Runa, we cross the Eyjafjarðará river, on an interesting rock bridge, upstream of which is a pretty pool. From the bridge, continue up a steep slope to the highlands.

Mt. Hvassafell beyond the valley Stóridalur in Eyjafjörður. Photo: Dagur Bragason.

At the bottom of the Eyjafjörður valleys. Photo: Jón G. Snæland.

An incredible and still memorable event took place starting at Runa, on September 27, 1898. On that day, three shepherds rode up Eyjafjarðardalur and intended to look for sheep above the shoulders of valley. They set off hiking up the valley in light clothes and without food. One of them, Kristinn Jónsson, went up the east side of the valley, planning to return down along the west side. When Kristinn reached the top, a thick fog covered the area and he became lost, and took a wrong turn, arriving at the river branch Bergvatnskvísl, which is the source of the Þjórsá, Iceland's longest river. Kristinn decided to walk along the river, and in so doing, arrived at another sizeable stream, which he waded across. Kristinn continued walking for eight days until he was found, in bad shape, at Búrfell, (in south-central Iceland). What saved Kristinn was that he decided not to cross Bergvatnskvísl. If he had done so, he would have ended up at Búðarháls at the site known as Hald, where the rivers Tungnaá and Þjórsá would have trapped him, and been impassable obstacles.

Once reaching the top of the mountains and entering the highlands, the road quickly goes south through the high pasture known as Nýjabæjarafrétt. Soon there is an intersection with the Dragavegur road, F881, coming from the east, which leads to the Sprengisandsleið track. Our route takes us to the right and ends at the hut run by the touring club Ferðafélag Eyfirðinga. Accommodation in the hut or at the campground can be bought here. There is a large, natural thermal pool here.

Fjallabaksleið nyrðri (Landmannaleið)

Start – Finish	On Route 1, at Eldvatn – At the intersection at Landmannalaugar
Distance/Time	55 km. Four to five hours' drive.
Fuel	At Kirkjubæjarklaustur and at the Hálendismiðstöð station at Hrauneyjar.
Caution needed due to	Steep slopes and many river fords.
Passable for	All jeeps and vehicles with high clearance during summer.
Of interest	Rivers, colourful mountains, waterfalls, and volcanoes.

We enter this route from the Ring Road, Route 1, in south Iceland at the river Eldvatn, and follow Route 208 north into the valley Skaftárdalur, to the farmsite Búland. This is the start of the actual mountain road, F208. From here, we continue in to the area known as Fjallabak (Behind the mountains) driving along Núpsheiði heath. Next we cross over the river Syðri-Ófæra on a bridge just south of Hólaskjól. The river is gorgeous here as it cascades through a large chasm under the bridge. To the right is the Skaftá River, a fast and threatening glacial river that has its source in Vatnajökull, Europe's largest glacier. By the roadside, Hólaskjól is visible, with a few buildings here. It is possible to pay for a stay in a small cottage here, or for a place to sleep in your own sleeping bags in the larger cabin. There is also a good campground. Upstream from the buildings at Hólaskjól, it is a

Mountains on the Fjallabaksleið route. Photo: Jón G. Snæland.

short walk to the Syðri-Ófæra river where it falls into a canyon, and where it is worthwhile having a look at a beautiful waterfall called Silfurfoss or, by some, Litli-Gullfoss.

Continuing inland past Hólaskjól, we arrive at the fording site across the branch Strangakvísl, which is quite a challenge for ordinary vehicles. Therefore, it is important for those driving such autos to thoroughly check the status of the stream before entering the river. We then drive up a steep

In Jökuldalir valley. Photo: Guðni Elísson.

An old, rather dangerous line road over Jökuldalakvísl below Mt. Stakafell. Photo: Jón G. Snæland.

hill where the road going to Eldgjá canyon meets our route. Eldgjá is the site of a major volcanic eruption in the year 934, along a 40 kilometre-long fissure, which is part of the same fissure system as famous Mt. Katla in Mýrdalsjökull. It was the largest eruption in Iceland's history and the lava from it covers 700 square kilometres. The Eldgjá canyon is 270 metres deep, and the Nyrðri-Ófæra river plunges into it in a beautiful waterfall known as Ófærufoss. The waterfall used to have a pretty natural rock bridge across it, but this collapsed sometime during the winter of 1993. The track into Eldgjá ends at a large parking place, and getting there requires crossing the Ófæra river in several places at fords suitable only for jeeps. From the parking area, it is possible to walk to the waterfall.

We continue above the turn-off leading to Eldgjá, and there's a steep drive up a hill onto the Herðubreiðarháls ridge, where there is a place to park and an excellent view. North of Herðubreiðarháls is an intersection and the branch to the right leads to Langisjór, a clear natural lake 20 kilometres long, surrounded by a beautiful ring of mountains. Many Icelanders consider Langisjór and its surroundings to be the most beautiful place in Iceland. The route to Langisjór is passable for all vehicles from Fjallabaksleið.

We go farther past this intersection and there is an optional track leading right/northeast to Faxasund, which is only passable for jeeps. We continue straight on and down into a lovely valley called Jökuldalir. Then unbridged rivers and creeks are forded, and it is vital that those in smaller vehicles examine conditions carefully before crossing these. It is also possible to drive along Jökuldalir to the north, where there are huts, including an old turf one called Glaðheimar. However, the north option has some deep fords to cross.

Finally, we cross a ford downstream from Kirkjufellsvatn lake, and drive up the hill that follows. Over the hill is one more optional track to the right, and this one goes to the hut known as Höll (Palace). This is passable by street vehicles, and it is pleasant to pause at the verdant slopes near the hut for a snack break. From there it is a short way to the intersection leading to Landmannalaugar, or onward via the Sigölduleið which leads north to Hrauneyjar.

Fjallabaksleið syðri

Start – Finish	At Keldur in the Rangá district – In Hvanngil canyon, north of Mýrdalsjökull glacier.
Distance/Time	62 kilometres. About three hours' drive.
Fuel	At Hvolsvöllur.
Caution needed due to	Many waterways that can be dangerous and impassable in wet weather; for example: Markarfljót, Grashagakvísl, Bratthálskvísl, Hvanngilslækur, Kaldaklofskvísl and many unnamed creeks.
Passable for	All jeeps, but not the smallest ones.
Of interest	Beautiful mountains, glaciers and rivers.

This track starts at the ancient farmsite of Keldur and is called Fjallabaksleið syðri (F210). The initial part traverses lavas known as Skógshraun and Langvíuhraun. The road through the lavas had been quite rough, but Mt. Hekla's eruption of the 2000 layered cinders on the track and made it easier going. The route goes inland along the river Eystri-Rangá, which runs through beautiful chasms. A rough section lies near the river at the site called Foss where there is an abandoned farm that has been converted to a public guest hut accommodation. On the other hand, when emerg-

The Álftaskarð pass to the left, and just a glimps of the lake Álftavatn. Mýrdalsjökull in the background. Photo: Jón G. Snæland.

ing from the lava the track improves, but there can be substantial runoff from Mt. Hafrafell in early summer, before the Roads Administration has graded it.

In the southern reaches of the Rangá bottoms, the track branches to the right leading to Hungurfit, which is discussed on page 86-89. But we go straight.

At Laufafell is an intersection where we continue south of the mountain. A bit farther past Laufafell, our track descends to a creek, named

The facilities for travellers at Hvanngil. There are two huts with sleeping-bag accommodation, as well as a camping ground and sanitary amenities. Photo: Jón G. Snæland.

The older round-up hut in Hvanngil. Photo: Jón G. Snæland.

Laufalækur by Oddi Oddsson, a well-known mountain-man of the Rangá district. Drive into the creek and turn directly left, upstream while in the streambed. Continue following the streambed about half a kilometre, until it bends and comes to a steep slope, where the exit track is distinct and it drive up it, as the track emerges to the right. At the top of the slope, it is ideal to pause and enjoy the view across the sandy expanse of Launfitarsandur, plus Ljósártungur and inland to Jökultungur under the mountain Háskerðingur, which is topped by the tiny glacier Kaldaklofsjökull.

From the slope, we drive down to the ford on the Markarfljót, the largest crossing on this route. The river bottom is coarse gravel, but fairly flat. Depth is usually between 60-70 centimetres, but it can be more during wet periods. Those travelling alone should wait until another vehicle is in sight before attempting to cross. Soon after crossing the Markarfljót, a smaller river is crossed, called the Ljósá. Then we go along a large high knoll with beautiful views to either side. Then we arrive at a mountain gap called Álftaskarð, from which there is a good view to Álftavatn lake, and the Ferðafélag Íslands hut at the north end of the lake.

Once emerging through the gap, an intersection is encountered. We do not take the right option as it leads to the lake and beyond, and is a hazardous route only for the highly experienced wilderness driver. However, taking the left option leads to the FÍ hut where it is possible to get the latest information on the area from the hut supervisor.

At the previously mentioned intersection there is a shallow cave in the mountainside, which was a dwelling place for shepherds, fishermen and hunters. Centuries ago, before swans were protected, hunters captured them during their molt at Álftavatn (which means 'Swan Drake Lake'). The cave was also used for storing boats. But a boat was probably not available in 1838 when Benedikt Erlingsson, a farmer from nearby Fljótshlíð, drowned while trying to swim out and capture molting swans.

There is often a lot of traffic in the area around Álftavatn, not least because this is where the popular hiking trail 'Laugavegurinn' goes along the route as far as Hvanngil. The hiking trail goes from Landmannalaugar down to Þórsmörk Preserve. From Álftavatn it is a short distance over to Hvanngil where Ferðafélag Íslands also has a guest hut, which is the end of this route.

Fjörður

Start – Finish	**North of Grýtubakki – Sea shore at Kaðalsstaðir.**
Distance/Time	**27.6 km. Four hours round trip.**
Fuel	**At Grenivík.**
Caution needed due to	**No real difficulties along the way.**
Passable for	**Jeeps and larger vehicles. Note there is a 3-ton weight limit on the bridge at Illagil chasm and bridge width is 2.45 metres.**
Of interest	**Abandoned farms, beautiful mountains and nature.**

We drive along the east side of the longest fjord in north Iceland, Eyjafjörður, and turn right/east off the main road just north of Grýtubakki and enter a road marked Hvalvatnsleið, F839. From there, via the Leirdalsheiði heath, it is 19 kilometres to the hut known as Gil, and this track is passable to all vehicles during summer. There are small brooks along the way to ford. The first one is Grýtá, which is crossed at a very fun ford. However, the largest stream along the way is the Gljúfurá, which is bridged, but there is also a good ford beneath the bridge, if drivers wish to try crossing at a ford.

The highest part of this route is Hávörður, which has an elevation of 350 metres above sea level, after which it descends along the so-called Sporður (Tail) which is a narrow slice of land between the canyons of Illagil and the river Gilsá to the west. At the lowest end of the slice is the hut Gil, which is supervised during summers, and there is a pedestrian bridge at the hut that crosses the Gilsá. Below the hut we cross Illagil on

Driving over Gljúfurá on the way to Hvalvatnsfjörður. Photo: Jón G. Snæland.

an old bridge that used to bridge the river Fnjóská. The bridge has a three-ton weight limit and it is only 2.45 metres wide. The route to the Gil hut is well posted and therefore can be driven by suitably equipped jeeps during the winter, but the slopes at Illagil are difficult in winter. If it is impassable

Cheerful travellers stop for a bite at the Gil hut. Photo: Jón G. Snæland.

Driving over the old bridge at Illagil. The bridge is narrow, only 2.45 meters, and note the 3 ton weight limit. Photo: Jón G. Snæland.

at the canyon, drive down the fjord on the west/left side. But note this route from Illagil out the fjord is not posted.

From Illagil canyon we drive down along the Hvalvatnsfjarðará river,

In Hvalvatnsfjörður. Photo: Jón G. Snæland.

Travellers looking around the spectacular beaches of Hvalvatnsfjörður fjord. Photo: Jón G. Snæland.

and along the way travellers should note the 13 abandoned farms at Hvalvatnsfjörður and Þorgeirsfjörður. At the roadside on the east we first pass Jórunnarstaðir, then Kussungsstaðir which is across the river. Finally, we pass Kaðalsstaðir, which was the outermost farm on the east, and now has a nice campground with sanitary facilities near the ruins. From Kaðalsstaðir we cross the river Kaðalá and approach the seashore, where the remains of boat landing sites are visible. To the west above the coast are also quite a few ruins, including those of the farms Arneyri and Brekka. It is generally impassable there during summers since the Fjarðará river has a heavy current at its mouth. It is absolutely impossible to cross and has claimed the lives of some of those trying.

Over 100 people lived in Fjörður and Þorgeirsfjörður in its prime, during the latter part of the 19th century, but the area became totally abandoned in 1944 when the matriarch of the area, named Hólmfríður, moved from Tindriðastaðir.

If travellers do not trust themselves to drive the route described in this section, it is possible to take a day trip under the guidance of Bergvin

Jóhannesson, who lives at the farm Áshóll (tel: +354-462-6914). He takes 4-20 passengers at a time.

The way over to Þorgeirsfjörður is for hikers only since the track has been closed to vehicle traffic. The trail there is three to four kilometres and the Fjarðará is bridged along the way.

Forsetavegur

Start – Finish	**Fjórðungsvatn, F26 – The Laugafell huts, Laugafellsskálarnir, F752.**
Distance/Time	**30.2 km. About an hour's drive.**
Fuel	**In Eyjafjörður, at Goðafoss waterfall and at Hrauneyjar.**
Caution needed due to	**Fording the stream branch Bergvatnskvísl.**
Passable for	**All jeeps can travel this route, also smaller ones, if water levels are low.**
Of interest	**Ancient cairns, desert and gorgeous mountain views.**

Forsetavegur means President's Road, and it draws its name from the fact that Kristján Eldjárn, as President of Iceland, once went to the north part of the country on an official visit and the route was graded and improved for that occasion. It is not clear whether he actually travelled that part of Sprengisandsleið, but nonetheless the presidential label has stuck to this section of road.

From Sprengisandsleið, F26, we drive onto the route west of Fjórðungsvatn lake. Then we cross over a small rise and past the turnoff leading to the south, which is called Biskupsleið. Once past Biskupsleið we go down

Hallgrímsvarða cairn. It was raised in honour of Hallgrímur Jónsson, author, teacher and long-time member of Ferðafélag Íslands. Photo: Jón G. Snæland.

to a low area, where a faint track to the north can be detected, at the base and to the east of some sand dunes. The track leads to an ancient cairn called Beinakerling, (Bone Crone) and it stands among 24 'sisters'. The cairn is 1,900 metres north of the main road (N 64 54.053-W 18 06.392) and driving here is prohibited. Beinakerling stands on a small hill, east of a higher dune, where old cairns marked the way across Sprengisandur, but the hill itself is not prominent in the landscape. Beinakerling was generally considered lost in the latter part of the previous century, when a doctor, Björn Jónsson, found it again in 1978. Mention is made in the book *Landið þitt Ísland* ("Your Country, Iceland") that in cairns such as this greetings have often been left behind for travellers who might follow, with a rhyme on the word 'kerling'. Some of these were frequently rather coarse and not really printable, although some survive. The greetings were stuffed into a shank bone of a sheep or cow, which was stored in the cairn itself, which explains the boney reference.

From Beinakerling, the route leads out into a vast sand plain, where it is believed the explosion (Sprengir) occurred from which the name Sprengisandur originates. The site name *Sprengir* is known as far back as 1476, but in the oldest manuscript from 1820–1830, from the judgement of the men from Holt and Landmannaafréttur, a quote says, "between the river Túná and Sprengisandur" (Icelandic Ancient Archives, VI:81). The plain Sprengir is believed by some to get its name because horses were ridden so fast through here that they nearly burst. Another story mentions that an

A group of travellers from Útivist touring club view the old cairn known as Beinakerling (Bone Crone) and 24 'sisters'. Photo: Jón G. Snæland

international airline pilot had marked Sprengir on his flight-books and felt it would be possible to land a passenger jet in an emergency, because the sands were so smooth.

The prominence called Útsýnisalda, which is 910 metres high and east of Háalda, was given its name in 1950 by the touring club Minnsta Ferðafélagið, who built a cairn there and encouraged travellers to drive to the top and enjoy the view. Those were different times, when it was considered normal to drive wherever possible, without concern for the environment. Members in Minnsta Ferðafélagið were the first to drive many of the tracks which are still in use today.

When we are west of Sprengir, we arrive at a small clear-water branch, called Bergvatnskvísl, which is the source of the Þjórsá river, Iceland's longest river. Bergvatnskvísl is a free stone stream, collecting tributaries along its way, with rather low volume during the summer, and should be passable by all jeeps on dry days. However, the river is quick to grow into a large, dangerous stream during rainfall, as it did in 1989 when it tragically claimed three lives. The Bergvatnskvísl watershed reaches all the way north

Here a jeep has landed in one of the warm streams west of Háöldur. Icelanders use super-jeeps on their frequent winter trips in the highlands. Photo: Jón G. Snæland.

to Dragaleið, and between Forsetavegur to the west and Sprengisandsleið to the east. This is a huge, vast area. West of the ford on Bergvatnskvísl, our way leads to the north and west of Háöldur. In this area there is a lot of geothermal warmth and the brooks that run from Háöldur to Hnjúkskvísl are a challenge. It is said that on the prominence of Háöldur there are small, natural warm pools.

From here our way goes north between Laugafell to the east and Laugafellshnjúkur to the west. It is possible to drive up on to Mt. Laugafell on a track on the northwestern slope of the mountain. The track is passable for all modified jeeps and it is certainly worth the effort to drive to the 904-metre summit. From here the view to the south spans all of Sprengisandur and past Hofsjökull glacier, and to the east the glacier, Tungnafellsjökull. From Laugafell it is a short drive to the Laugafellsskáli hut. But first, there is a ford across the branch Laugakvísl, a small brook in summer, but it can be an obstacle during winter due to thermal heat melting ice and snow that normally bridge the brook.

Gæsavatnaleið (Vatnajökulsvegur)

Start – Finish	Sprengisandsleið F26 – Dreki at Dyngjufjöll mountains.
Distance/Time	104 km. Six to eight hours' drive.
Fuel	At the Hálendismiðstöðin in Hrauneyjar and at Reynihlíð at Lake Mývatn.
Caution needed due to	Glacier rivers, wet sands and large boulders.
Passable for	Jeeps, but not solo. Preferably several vehicles together.
Of interest	Glaciers, lava formations, deserts and mountain views.

We turn into Route F910 off Sprengisandsleið, F26, north of the glacier Tungnafellsjökull. The road north of Tungnafellsjökull is passable and easy going, though glacial rivers can be obstacles. About midway on this route we cross a stream called Jökulfall, whose source is in Tungnafellsjökull. Jökulfall can be quite an obstacle for smaller vehicles in the latter part of warm, summer days. Early in the morning on the other hand, volume is much less and crossing it is no problem. The glacial river, Langadrag is a bit to the east and much easier to cross. East of Langadrag is an intersection where it is nice to turn south in the direction of Vonarskarð. After about 2.6 kilometres, we take the track east to the left which goes to the river Skjálfandafljót. There is a majestic waterfall here, called Gjallandi,

Travellers at Hitulaug. Photo: Jón G. Snæland.

The flood-plains of the river Jökulsá á Fjöllum, below the glacier Dyngjujökull. The black mountains beyond the jeeps is in fact the glacier, covered with sand and mud. Photo: Jón G. Snæland.

and from here it is possible to drive down along the river and back to Route F910 again, at a bridge over the river. East of the bridge on Skjálfandafljót is an intersection. To the left (north), Route F910 goes on to Ódáðahraun, and to the right, or south, are the Gæsavötn lakes. In the old days, the route north of Vatnajökull was not often mentioned, but known as Vatnajökuls-

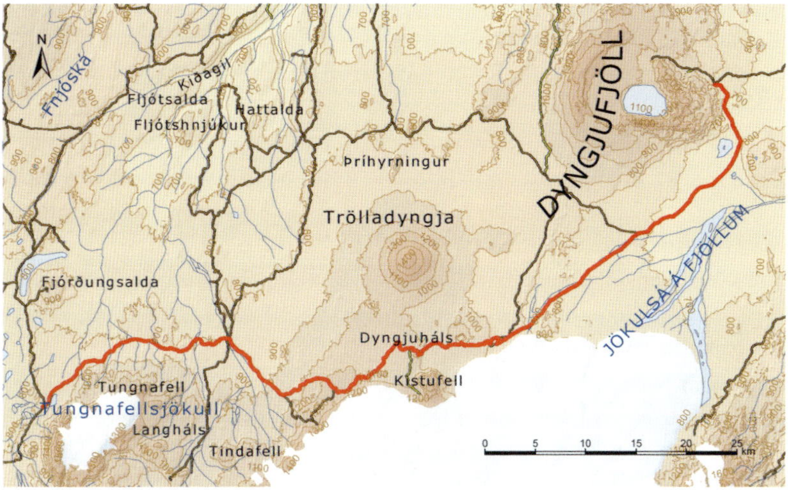

Vörðukambur (Cairn Ridge) on Gæsavatnsleið, where travellers have raised numerous cairns. Photo Jón G. Snæland

vegur. Members of the group Minnsta ferðafélagið explored the route in 1950 from the west, but did not make it all the way, stopping 12 kilometres east of the Gæsavötn lakes. Next they approached from the east in 1951, finally making it all the way.

Gæsavötn were found in 1880 when people from the Mývatn area and the valley of Bárðardalur were exploring southward to the glacier, Vatnajökull. There is a nice hut at Gæsavötn, called Gæsavatnaskáli, which is privately owned by the group Gæsavatnafélagið.

Our way continues past Dyngjuháls and there is a collection of cairns at the place known as Vörðukambur. This is the highest stretch of road in Iceland, with an elevation of 1240 metres. East of Dyngjuháls we see the emergency shelter Kistufell, which is on the east slope. The hut is open and it is popular place to eat one's snack or meal. South of the shelter, on the fringe of Vatnajökull glacier is Mt. Kistufell, 1,444 metres high. Kistufell made the news in 1950 as the site of one of the most famous passenger aircraft crashes in Iceland. An international flight involving a plane named *Geysir* crash-landed on Vatnajökull. All on board survived, but the wreckage is still on the glacier.

East of Kistufell we arrive at Urðarháls, a very rocky section where the track is not clear. Urðarháls is a shield volcano from a warm interval in the Ice Age and has Iceland's largest shield volcanic crater, about 110 me-

A beautiful waterfall in Hraunkvísl, not far from The Gjallandi waterfall in Skjálfandafljót river. Photo: Jón G. Snæland.

tres deep, 1100 metres long and 800 metres wide. East of Urðarháls, the branches that collectively make the glacial river Jökulsá á Fjöllum are encountered. We drive down the slopes of Urðarháls onto a large sandy plain. Southward, to the right, Dyngjujökull is seen in the distance, often dark from its own sediments. Late summer afternoons a flood wave comes from the glacier, flooding the sandy plain for a width of one to two kilometres, but the water is rather shallow. The way across the sands is posted, but if the surface is wet, it is best to drive in low four-wheel-drive gear and maintain a

The Skjálfandafljót river gorge upriver from the rod bridge. Photo: Jón G. Snæland.

steady speed until the crossing is complete. DO NOT STOP as the vehicle may quickly become entrenched in the sand, and it is a major operation to remove stuck vehicles. Those who do not trust themselves to cross the sands can cross via a detour on Route F910.

Intersections in the highlands are unclear, but the start of the highland track is approximately at the GPS quadrants: N64 49.513 W17 04.217 and the next point above the flood plain Flæðurnar is N64 49.656 W17 02.572, by which time we are pretty much safely on the way across. Once the flood plain is crossed, Route F910 is taken and becomes soft black sand and we head north of the prominence Vaðalda. At Vaðalda there is a sign with directions to the river Svartá, which is a pleasant place to visit. The Svartá is a sizeable spring-fed stream that emerges full-fledged from coal-black sands. It then runs a short distance to join Jökulsá á Fjöllum in a beautiful waterfall named Skínandi, meaning "Shining". From Vaðalda, it is a short distance inland to the west to the Ferðafélag Íslands information centre at Drekagil.

Hagavatnsvegur

Start – Finish	**At the Sandá river on Kjalvegur, F35 – Hagavatn lake.**
Distance/Time	**16.7 km. 45-60 minutes' drive.**
Fuel	**At Geysir in Haukadalur.**
Caution needed due to	**Steep slopes and glacial streams.**
Passable for	**All jeeps, also smaller ones.**
Of interest	**Glacier lagoon, glaciers, beautiful mountain scenery and waterfalls.**

Hagavatnsvegur, Route F335, starts at Kjalvegur north of Gullfoss waterfall and then we turn west along the Sandá river in a flat area to the Sandvatnshlíð slopes. Here we go uphill above Sandvatn lake. This route has often been driven in small autos during the summer under the best conditions as far as Einifell, beyond which it is rather rough for small vehicles. In the process, we pass the turnoff down to Sandvatn lake. Here experiments are being conducted to try to stop wind erosion on the small patches of vegetation in the barrens north of the lake.

When approaching Einifell, mountain peaks appear in a line stretching north along the glacier Langjökull. This row of peaks is called Jarlhettur, or the Earl's Hats. Further north, off another side road, and near Skálpanes, a tour company has established a base camp for staging snowmobile trips from there onto Langjökull. To get to their base just past Skálpanes, go further north on Kjalvegur, up to Bláfellsháls, turning left/west on Route F336, which is quite passable and starts opposite Bláfell.

There is a small guest

Nýifoss waterfall where the river Farið flows from the glacial lagoon Hagavatn. Photo: Jón G. Snæland.

hut run by Ferðafélag Íslands at Einifell, and it accommodates about 12 persons. It was built in 1942 and has since been twice renovated. The hut stands in a pretty, but small grassy hollow north of Einifell, just east of the glacial river called Far. There is a lot to see within a short distance of the hut. For example, it is a short walk to Hagavatn lake or in to the Jarlhettur (Earl's Hats). But one cannot go further than to the hut in small vehicles (and only that far with care in the best conditions). Those in jeeps can cross a harmless glacial branch, called Jarlhettukvísl, to get to Hagavatn. Follow a rough jeep track westward to the glacial river Far. This river drains from Hagavatn, in a canyon with falls and rapids, dropping 100 vertical metres from the lake to the surrounding flats. The waterfall is called Nýifoss (New Waterfall). A very powerful jeep in low gear is required to get up the very steep slopes all the way to Hagavatn lake. Others simply walk up the slopes, where an amazing view awaits. Hagavatn is a glacial lagoon, and the glacier Eystri-Hagafellsjökull, which branches off larger Langjökull, plunges into the lagoon. Previously, a pedestrian bridge spanned the river Far, but when the Eystri-Hagafellsjökull burst forward a full kilometre in

Nýifoss waterfall in the river Farið. Photo: Jón G. Snæland.

1999, the Far flooded, taking the bridge with it. The only trace left is a buttress on the north bank.

The Ferðafélag Íslands Annals for 1980 comments:

> At one point there is a small nook in Brekknafjall, and below it a narrow, dark line runs down the gray rocky slope. When approached more closely, it becomes evident that this is a canyon. Those standing on the gravels opposite the opening see only a black abyss when they look up the canyon. This is Leynifossgljúfur (Hidden Falls Canyon) where the Far used to flow during the years from 1929 to 1939. In 1929 Hagavatn burst forward and the Far flowed here, forming the waterfall Leynifoss, until another natural glacial moraine dam failed 10 years later, giving the Far a new canyon, with a new waterfall now called Nýifoss (New Waterfall).

Haukadalur and Haukadalsskarð

Start – Finish	**Haukadalur valley in Dalasýsla district – Brú in Hrútafjörður.**
Distance/Time	**31.9 km of which 17 km are highlands. One and a half hour's drive.**
Fuel	**In the town of Búðardalur and at Brú in Hrútafjörður.**
Caution needed due to	**One shallow ford on the Haukadalsá river.**
Passable for	**All jeeps, but it is slow-going.**
Of interest	**Museum at Eiríksstaðir rauða, abandoned farms, pretty mountains and waterfalls.**

Off of Route 60 in west Iceland, we turn northeast into the valley Haukadalur on Route 586, and proceed to the lake Haukadalsvatn, and the farm Stóra-Vatnshorn. This is the site called Eiríksstaðir, which is a replica of the farm of Eiríkur rauði (Erik the Red) Thorvaldsson, who first settled land in Greenland about 985 AD. Erikur's son, Leifur Eiríksson, Leif (sic) the Lucky, was born at Eiríksstaðir and moved as a young boy with his

Well-built cairns from times past along the track over Haukadalsskarð pass. This was formerly a much-travelled route and it was important for travellers to follow the cairns so as not to lose their way, since it could be fatal to get lost in the mountains, whether walking or riding. Photo: Kjartan Gunnsteinsson.

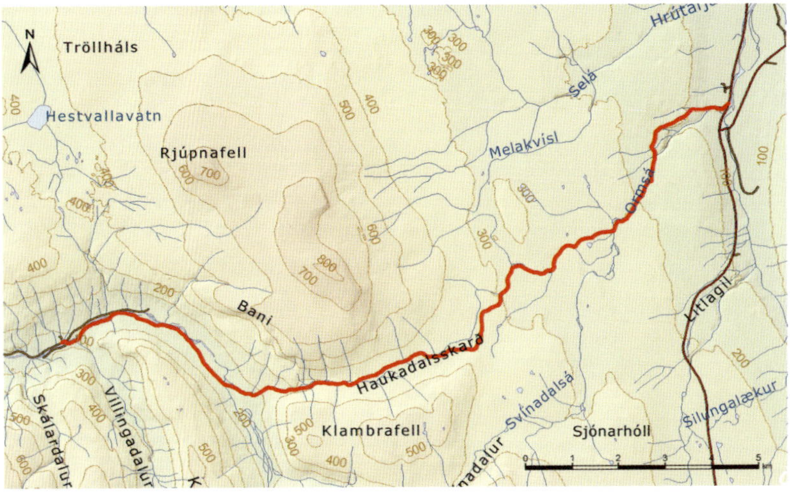

father to Greenland. Leifur heard of land west of Greenland and sailed there along with about 30 people in the year 1000. He discovered North America, specifically landing at Baffin Island and Labrador, and the land, which Leifur called Vinland, was in all likelihood where Newfoundland is

Travellers by the old sheep corral deep in Haukadalur, a few hundred meters above the ford in the Haukadalsá river. Icelandic farmers keep their sheep in common land in mountains and moors over the summer. In the autumn they herd them into corrals where each farmer has to pull his own sheep from the main corral into a separate partitiom, from which he then drives them back to his farm. Photo: Kjartan Gunnsteinsson.

Norðlingafoss waterfall in the river Ormsá, which was harnessed, providing electricity to the Brú telephone station, among others. Photo: Kjartan Gunnsteinsson.

today. In Vinland, Leifur and his shipmates built several houses and stayed the winter. On their way back to Greenland, Leifur rescued 15 shipwreck survivors who were clinging to a skerry and received the title, 'The Lucky'. At Eiríksstaðir, visitors travel back 1,000 years in time, meeting descendants of Eiríkur, Leifur and other Vikings, in authentic Viking costume.

There are many other historically significant places in Haukadalur. Among them is Jörfi, where an autumn celebration, *Jörfagleði*, was held for ages until banned by official decree in 1708. One explanation was that 19 illegitimate children were born following the Jörfagleði a year earlier.

Deeper up the valley, the river Haukadalsá is crossed on a bridge and we emerge on the south side of the valley, where the actual mountain road Route F586 begins. Drive still deeper up the valley, past a round-up corral built in 1961, and on the north slope of the valley are several abandoned farms. Up the valley, the place known as Jasonarstaðir is visible as is a summer cottage. Further up the valley from this is an intersection where we turn left and down a slope to a ford on the Haukadalsá. The Haukadalsá is rather small during summer and crossing is straight-forward, although the bottom is somewhat coarse and rocky. Above the ford a pretty waterfall is visible; it is called Réttarfoss and can be reached by taking the right turn at the previously mentioned intersection.

There are ruins of an old, dry-stone sheep corral just beyond the ford, and this is a comfortable place to pause for a snack and view both the waterfall and corral. Above the corral, the route goes through a pass between the mountains, Klambrafell to the south and Geldingafell to the north.

The pass, Haukadalsskarð, is an old cross-country route, where people travelled on foot between the north and west country. During winter this proved extremely arduous travel in blizzards and drifting snow, and many perished in the effort. Once we reach the livestock quarantine fence, where the route is highest above sea level, the landscape then begins to descend towards the Hrútafjörður fjord. Soon the track becomes vague and at one point we must follow the streambed of a rough brook for a short distance. Below, we arrive at a dam on the river Ormsá, where the route follows the edge of the canyon Ormsárgil. Eventually, we drive across the river at a shallow ford. Below the ford, in quite an impressive canyon, is the waterfall Norðlingafoss on the Ormsá. We continue once again along the canyon rim of the Ormsá, nearly all the way to Route 61, where our journey ends at Hrútafjörður.

Hjörleifshöfði

Start – Finish	**On Route 1 – Ring Road, north of Cape Hjörleifshöfði.**
Distance/Time	**8.2 km. Allow for 1-2 hours.**
Fuel	**At the town of Vík í Mýrdal.**
Caution needed due to	**Wet sands in the stream on the east of Hjörleifshöfði.**
Passable for	**All jeeps, though not solo.**
Of interest	**Farm ruins, caves and beautiful cliffs.**

Hjörleifshöfði is a headland, or cape, on Mýrdalssandur east of the village of Vík í Mýrdal. When we drive over the stream of Múlakvísl and on to Mýrdalssandur, Hjörleifshöfði is seen on the right, rising high below Route One. We turn south/right off the Ring Road onto a good gravel road, just north of the headland that then sweeps west of it. Bluffs, named Rituberg, protrude on the west edge of Hjörleifshöfði, and the main hiking path to the top lies in a notch in the bluffs.

Driving around Rituberg to the south brings us to the emergency shelter of the local rescue team Víkverjar from Vík í Mýrdal. The hut is an iron-clad, timber structure, which can house 12 persons. Just inland from the shelter is a large open cave in the cliff. There are benches are here and a fireplace on the cave floor.

There is a large, interesting cave at the southern end of Hjörleifshöfði, where they have put in benches and made a fireplace in the middle of the cave floor. Photo: Jón G. Snæland.

But before going further, it might be fun to drive down to the shoreline, if the tide is right. It is about 2.5 km down to the beach along a sandy road, which can be a bit tricky for the smallest jeeps if the sand is dry. On the beach it is fascinating to follow the crashing of huge waves coming in from the open Atlantic. It can prove difficult to drive along the beach to the east or west because to the east the stream Mosalandskvísl is an obstacle, and to the west another stream, Miðkvísl, blocks traffic. These streams are not dangerous due to their volume, but rather due to the fact that they often hide quicksand in their beds.

The face of the "Cliff-dweller" can be seen from GPS coordinates N63 25.066 W18 43.953. Photo: Jón G. Snæland.

A group of jeeps crossing a small stream north of Hjörleifshöfði headland. Photo: Jón G. Snæland.

From the cave, those in larger 4x4 vehicles can continue around the headland to the east, and circle it. We drive between Hjörleifshöfði and impressive offshore rock pillars called Arnardrangur and Lásadrangur. The author encourages travellers to study the cliffs well, because at one point there appears to be a face pushing out of the rocks.

North of the headland it is necessary to cross the stream Mosalandskvísl at a coarse ford, which has a good bottom and a depth of 20-30 centimetres. The stream is again crossed somewhat further west, but here there is more chance of wet sands, so it is safer to cross with another vehicle nearby in case the vehicle becomes stuck. The west bank here is a high sand bank which is not passable for smaller jeeps. Beyond the sand bank, the circle is closed and the route ends.

Hjörleifshöfði is 221 metres high and made of volcanic tuff, or palagonite, which in Icelandic is called móberg, and therefore the headland is often called *Móbergseyja*, or 'Palagonite Island.' In the 14th century, the sea reached right to the base of the headland, but today it is almost three kilometres to the shore, which is Iceland's southernmost point and bears the name *Kötlutangi*.

Hjörleifshöfði is named for Hjörleifur Hróðmarsson, foster brother of Ingólfur Arnarsson, founder of Reykjavík, and one of Iceland's earliest settlers. Hjörleifur landed here and built two huts, but did not survive long because his slaves mutinied and killed him. Hjörleifur was buried at the highest point on the cape, which is now called Hjörleifshaugur. Also buried there is Markús Loftsson, his wife Málfríður Andrésdóttir, a brother and a child of Markús. Markús lived on Hjörleifshöfði from 1828 to 1906. There was a long history of settlement here, until 1721 when Katla erupted causing jökulhlaup (glacial flood burst) which destroyed the farm on the sands west of Hjörleifshöfði. The farm was relocated higher up on the headland, but abandoned in 1936.

Hrafntinnusker

Start – Finish	**Keldur – Hrafntinnusker (F210). Landmannaleið – Dalamót (F225).**
Distance /Time	**Kelduleið track is 60 km. but the Landmannaleið track is 9.8 km. Allow three to four hours, depending on choice of route and conditions.**
Fuel	**At the village of Hella and at Leirubakki in the Landsveit district.**
Caution needed due to	**Steep slopes, fords and snow drifts.**
Passable for	**All jeeps with caution, during summers.**
Of interest	**Mountain view, geothermal sites, ice caves and glaciers.**

There are two routes to the incredible area of Hrafntinnusker, either up from Landmannaleið (F225) or Fjallabaksleið syðri (F210). We will discuss both options, starting with Fjallabaksleið syðri.

The south approach starts along Route 264, north to the farm of Keldur, from Route 1 between the villages of Hella and Hvolsvöllur. From Keldur, we enter Skógshraun lava field, then go past Hafrafell mountain and enter another lava area, Langvíuhraun.

One of the most popular walking routes in Iceland lies between Landmannalaugar and Þórsmörk. It is called Laugavegur which also happens to be the name of Reykjavik's main shopping street. It reaches a height of over 1000 meters, crossing large spreads of snow, beautiful deep gorges and many rivers and streams. Photo: Kjartan Gunnsteinsson.

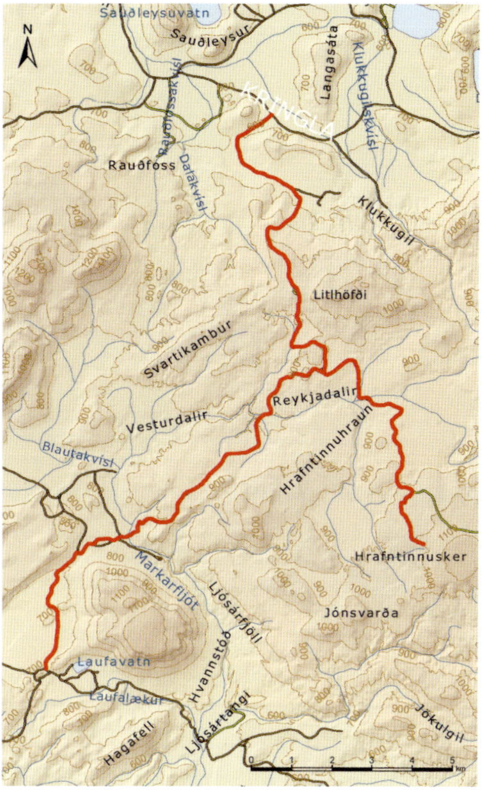

It is said that Neil Armstrong and other astronauts destined to land on the moon in 1969 held practice sessions in this area. Indeed, this is a lunar landscape with vast sands and lava fields, but as we approach the Grasleysufjöll mountains, the majestic glacier Tindfjallajökull appears.

At the intersection at Laufafell, turn left and head due north, skirting Mt. Laufafell, and arrive at the ford across the upper reaches of the river Markarfljót. Before crossing, it is worth driving a bit downstream to visit a pretty waterfall generally called Nafnlausifoss (Nameless Falls). Not all agree with this and claim it has long been called Lýsingur. Others feel it should be named for Rudolf Stolzenwald, the mountain-man who built the Dalakofi hut just above the ford in 1971.

From the falls, we cross the ford on the Markarfljót and drive up the steep slope onto Reykjadalsleið track. At the top of the slope, the valleys of Reykjadalir appear in all their splendour, and we seem to be in another world. Green moss dominates everything and the Hrafntinnuhraun lava rises from a rugged landscape. At the Reykjadalir valleys, Markarfljót is again crossed, but this ford is shallower.

The Reykjadalir valleys have a tremendous amount of geothermal energy and steam billows up widely. The route through the valleys often involves steep slopes and snow drifts are frequently seen well into summer. Above the valleys is an intersection known as Dalamót. Take a right turn here and cross a small river, and up a slope towards Hrafntinnusker, a vast ridge of volcanic glass, obsidian. At the first slope onto Hrafntinnusker it-

The ice cave at Hrafntinnusker. Photo: Jón G. Snæland.

self, it is possible to stop at a small hollow on the right and view an ice cave melted into a huge snow drift by a thermal vent. Here thermal hot spring water blends with a cool brook to form a lukewarm "puddle".

On the slope ahead the route divides, with the left branch for super-jeeps only, leading up a steep slope to a hikers' hut, Höskuldarskáli, north of and at the base of Hrafntinnusker. Super-jeeps are required because this track crosses the ice of a glacier tongue. However, the other branch leads straight ahead to a large ice cave at the margin of Hrafntinnusker. The cave has suffered from the effects of global warming, and extreme caution should be observed if entering the cave, because the roof has often caved in, sometimes causing loss of life. The mountain named Hrafntinnusker is 1,128 metres high and was previously covered by much more ice.

Hikers can often be seen on the trail from Hrafntinnusker, passing the ice cave on their way along Iceland's popular hiking trail, Laugavegurinn, that draws its name from Landmannalaugar.

As previously mentioned in this section, it is also possible to reach Hrafntinnusker from Landmannaleið (F225). In doing so, we follow along the mountain Sátubarn, and up a very steep and eroded slope to a point opposite Mógilshöfði. On reaching the top of the ridge known as Pokahryggur, it is possible to climb a small butte called Þúsund metra höll (Thousand

Geothermal heat is widely found in Hrafntinnusker and the Reykjadalir valleys, as can be seen by numerous clouds of steam. Photo: Jón G. Snæland.

metre Castle), which, as the name implies, has an elevation of 1,000 metres above sea level. The view from here is spectacular with snow-capped mountains and glaciers in all directions. The list of glaciers and summits is impressive: Eyjafjallajökull, Tindfjallajökull, Mýrdalsjökull, Hrafntinnusker, Reykjafjöll, Kaldaklofsfjöll, Torfajökull, Vatnajökull, Hofsjökull, Langjökull and Eiríksjökull. The name, Pokahryggur, is rather new and comes from sand bags used by a man to fill gullies in the road, so he could gather obsidian for surfacing the front of the National Theatre in Reykjavík.

From Pokahryggur, the descent is along steep slopes, some extremely steep. There are frequently snow drifts here, well into the summer, and this can make navigating the slopes difficult. Down the last slope we enter a cluster of valleys and arrive at the previously mentioned intersection, called Dalamót, where we keep to the left to reach Hrafntinnusker.

Hrunamannaafréttur

Start – Finish	**Mt. Tungufell – Kerlingarfjöll mountains.**
Distance/Time	**75 km. Three to four hours' drive.**
Fuel	**At the village of Flúðir.**
Caution needed due to	**Numerous fords.**
Passable for	**All jeeps, but not smaller ones or SUV's.**
Of interest	**Mountain views, turf huts, and canyons.**

Driving north from the village of Flúðir, we take Route 30 and then turn right/northeast onto Route 349, and pass the church for the people of Tungufell. The church was built in 1860 and is now protected on the historical registry. At Tungufell, the highland track begins with a climb up a slope until we enter a pretty valley, Tungufellsdalur. Rising out of Tungufellsdalur, we again take a steep slope and soon arrive at an optional track leading to the left towards the Hvítá river. From the end of this track it is possible to walk two kilometres to the famous waterfall Gullfoss and see it from an unusual eastern perspective, as nearly everybody sees Gullfoss from the west side. It is also possible to take a small loop along the way and visit the ruins of the Hamarsholt that was abandoned in 1875.

Travelling further north on our main track, we arrive at another optional track, this one to the right/east. It is a line road called Tangaleið, following

The Svínárnes hut in the Hrunamannaafréttur, Mt. Bláfell in the background. The turf building this side of the hut is the old shepherd's shelter which has been renovated. Photo: Bjarni Kristinsson.

a Landsnet high-tension power line, and is for larger jeeps only. It crosses Sprengisandsvegur at the Sultartangavirkjun hydro power plant. There is much to see on Tangaleið, but due to many river fords, it is not passable for smaller jeeps. Among the first attractions worth mention is the canyon of the river Stóra Laxá, called Laxárgljúfur. At the east end of Tangaleið, after successfully achieving the crossing of the rivers Stóra Laxá and Fossá, are the high waterfalls of Háifoss and Granni.

Many rivers lace the area of the Hrunamannaafréttur high pastures, but one of the first on Hrunamannaafréttur is Melrakkaá beyond which is a posted track to the left down to the Búðarárfoss waterfall, a pretty waterfall on the river Búðará worth a look when in the area.

Continuing on our main northerly route, we first cross the stream of Blákvísl and then the small river, Búðará, where few stop. But, once, the Environmental Committee of the touring club 4x4 stopped longer than intended at the clear-water Búðará. This was during an early thaw and torrential rains in March of 2004. A large part of the group of super-jeeps was unable to cross over the river which had become a raging torrent. A few barely managed to cross but the rest had to wait 24 hours for the river to subside.

Just beyond the Búðará crossing is the area known as Rofshólar, where a hut of the same name, or formerly Rógshóll, used to stand. A jeep track was made to Rofshólar around 1957, but a jeep track further into the high pasture was not passable until the 1970's.

A group of travellers from the 4x4 Travel Club at Svínárnesskáli hut on their way to their club's hut, "Setrið" up under Hofsjökull glacier. Photo: Bjarni Kristinsson.

From Búrfell we can see inland towards the area known as Svínárnes, and on the lowlands between Búrfell and the river Hvítá a track is visible towards the Hvítá. This track is called Borarvegur and dates from 1960, when there were ideas of damming the river.

But now we are nearing Svínárnes where there are a few buildings that serve the district of Hrunamannahreppur. First we must cross the Svíná river, at a ford that is obvious, and we drive directly across on a flat sand bottom. The depth of the water is about 40 centimetres (16 inches). Svínárnes is a good place to take a snack or meal break, and it is possible to take a little time and walk over to the turf house built in 1956 on the remains of an old hut, which is next door to the main hut. Recently, the hut was renovated by the club called Ársæll that has worked for the restoration of old huts on the highlands. Sources tell of ruins of two other huts at Svínárnes that stood north of the current huts. There is a third hut across the river on a gravel ridge and it was made of stone. The hut from 1956 was mainly intended as a shepherds' shelter. There are remains of two other older huts at Svínárnes.

North of Svínárnes we arrive at the Rauðá river, which has large boulders in it, but little water during summer. When we have crossed the Rauðá, we continue through a flat, gravelly area along the Innri Rauðá river and approach the swell called Mosalda. At Mosalda there is an intersection where an optional right turn leads to the highlands and is called Klakksleið. If Klakksleið is taken it is possible to drive south and counter-clockwise east of the Kerlingarfjöll Mountains, and meet Illahraunsleið. From the intersection, Illahraunsleið can be followed to its end at the huts Ásgarðskálar, discussed in the section on Ásgarðsleið (pg. 21).

At the previously mentioned intersection at Mosalda, we continue straight and the route skirts various prominences along the slopes of Rauðárhlíð until we arrive at the gravel flats of the rivers Fúlá and Kerlingará that join just below, forming the river Sandá. The rivers are crossed at fords, which are often posted during summers. Beyond the rivers is the hut Leppistunguskáli, and in going there we cross the Kerlingará again. Kerlingará runs below the hut in a deep canyon said to be as much as 200 metres deep. A man named Jón Jónsson fell into the canyon about the turn of the century 1800, and survived. He fell into a deep pool and that is what saved him. From the hut we drive over the Kerlingará and take a right at the intersection.

We continue driving west of the swell Kerlingaralda, across the brook Stórilækur (Miklumýrarlækur) north of the swell and across two fords which are harmless during summer. Next we ford the stream Grjótá, and a few dozen metres north of the ford is an optional section for modified jeeps about four kilometres long, splitting off to the left leading down to Hænsnaver marsh and later to Fosslækjarver marsh. This track is fairly passable for modified jeeps until it nears Fosslækjarver, where it becomes more or less a dirt track that can be very difficult in the last kilometres, if not totally impassable in wet weather. At Fosslækjarver there is a nice turf hut, newly restored by the club Ársæll.

From the intersection mentioned above, we continue up a steep slope above the flats of Kerlingarflatir, up a rise west of the butte Kúpa. Beyond the rise is another intersection where an optional right turn leads towards Mt. Skeljafell where the track ends. From there it is a short hike to the cliff Kerling at the base of Mt. Tindur. Kerling is a 25 metre-high spire of palagonite that rises above a rhyolite talus slope. The Kerlingarfjöll mountains get their name from this spire. According to an old legend, the spire was formed when an old troll woman (tröllkerling) died here and turned to stone.

From the Skeljafell intersection, we skirt Mt. Skeljafell and follow along the river Fremri Ásgarðsá, which is a small clear-water river, from which it is easy going in to Ásgarðsleið.

Hungurfit

Start – Finish	**At the bridge over Markarfljót in Emstrur – South of Grasleysufjöll mountain on F210.**
Distance/Time	**28 km. Two hours' drive.**
Fuel	**At the town of Hvolsvöllur.**
Caution needed due to	**Glacial river Hvítmaga, very steep slopes and deep runoffs.**
Passable for	**Jeeps during summers, but not solo or smaller 4x4's.**
Of interest	**Glaciers, caves, and colourful mountains.**

From the village of Hvolsvöllur, we enter the valley of Fljótshlíð on Route 261 going left/north to Emstruleið, and turn left on the approach to and above the bridge over Markarfljót. The track should be visible to the left, but it may be difficult to see in early summer when few have travelled it. Once this track has been found, it is followed to a small chasm, which is entered to the right. This chasm is often an obstacle due to snow drifts that tend to persist well into the summer, in addition to steep side slopes, regardless of which side is chosen. From the chasm, the track is obvious down to the canyon Þverárgil.

It is a steep descent entering the canyon of the river Þverá, but once entered we drive as many as 18 times across the Þverá, at fords which are harmless clear-water sites in summer. Eventually, we arrive at an intersec-

Mt. Hattfell (Hattafell) seen from Þverárgil. Photo: Ingvi Þór Ragnarsson.

tion, where the left option is taken, although those in the largest super jeeps can take the right turn and take a circle drive along the north part of Þverárgil.

On the other hand, taking the left turn leads us soon to a narrowing in the route, where rocks have fallen from slopes on either side. It is so narrow

Entering Þverárgil involves careful negotiation as you drive between bolders that have fallen from the slopes above. The largest vehicles have to make a detour to enter the gorge elsewhere. Photo: Jón G. Snæland.

here that the largest jeeps have to squeeze along. At the other end of the narrows, we reach a grassy valley where it is ideal to pitch a tent. Across the 'campground' are noteworthy caves in the cliffs on the east side. In the old days, shepherds pitched their tents on the flats of the Þverárgil and also stayed in the caves, if necessary, but also used the caves for their livestock. One cave could hold 12 persons at once, but it was never a comfortable place to stay.

When leaving Þverárgil and proceeding on to Hungurfit, we drive up a very steep slope between the mountains of Lifrarfjöll and Litla-Grænafjall. In wet weather, this slope can be impassable to all but the best equipped jeeps.

The track down to the glacial river Hvítmaga, which has its source in the eastern slopes of the glacier Tindfjallajökull, is often so eroded that it is tough going for unmodified jeeps. The river is crossed at a ford, which can also prove an obstacle for standard jeeps, so it pays for travellers to examine the ford carefully before entering the river.

Once the river is crossed, we drive up onto a small ridge and to an intersection. Take the right turn, which leads quickly down to the site of Krókur and Krókskofi, a hut for the shepherds of the Rangá region. Here it might be nice to stop and view the landscape or eat a snack or picnic lunch. After returning to the intersection, the route continues easily to the turn-off to Hungurfitaskáli hut, although several small brooks are forded along

A group of jeeps on their way across the glacial river Hvítamaga, which can sometimes be a serious obstacle for smaller jeeps. Photo: Jón G. Snæland.

A low and narrow door closes the old cave that was used during the sheep roundup. Photo: Jón G. Snæland.

the way. From the hut, the way goes through Hungurskarð gap and across the river Eystri-Rangá to Fjallabaksleið syðri (F210).

In the Hungurskarð gap there is a unique cliff near the road. Those men who were making their first mountain trip had to climb the cliff at a certain point in their shoes made of skin and wearing mittens on their hands. Those who could not make the climb were not considered strong enough for a mountain journey. It proved difficult for many, even though it is not higher than three men's combined heights, because of the overhanging slant. This challenge was more in fun than something strictly enforced.

Innstidalur (Hellisheiði)

Start – Finish	Route 1, Ring Road-through the heath of Hellisheiði.
Distance/Time	20.5 km. Two hours.
Fuel	In the town of Hveragerði.
Caution needed due to	deep pools in rivers, steep slopes and pot holes.
Passable for	All jeeps, but not small ones.
Of interest	Hellisheiðarvirkjun Power Plant, hot springs and beautiful clear streams.

Although there are two tracks from which to choose when driving in to the valley Innstidalur, this is a description of the one farthest east of Reykjavík. The west route splits off Route 1, above the ski lodge at Hveradalir, going to the north. There we drive a semi-circle around the lava field Orrustuhólahraun and beneath the south slopes of Skarðsmýrarfjall. Beneath the mountain are also three scout huts, and it requires driving up a steep slope to reach two of them: those known as Bæli and Kútur. Kútur is the brown one to the east, and the other is Bæli. Both of these were damaged in the earthquakes of the year 2000 and are in the process of renovation. The third hut, Þrymheimar, is farther east and below the slope and is managed by the scout troop Landnemar.

Directions for walking routes at Sleggjubeinaskarð pass. There is a great view from the pass, including the new Hellisheiði geothermal power plant. Photo: Jón G. Snæland.

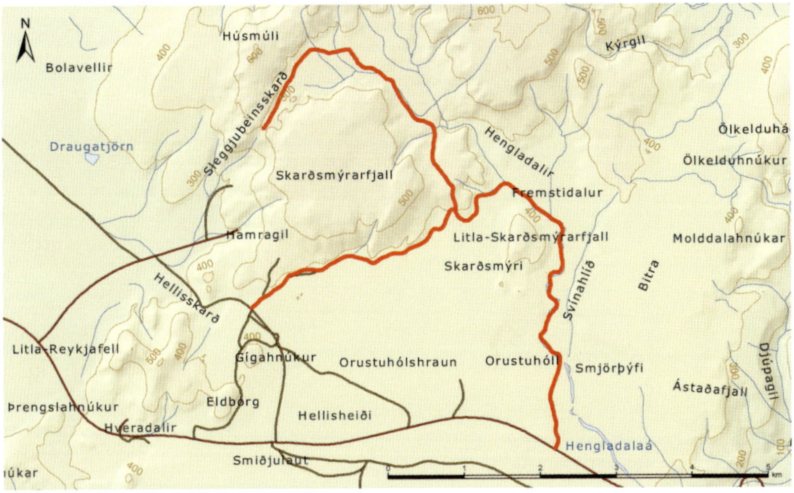

Just east of Þrymheimar is an intersection from which our route to Innstadalur actually begins.

We drive north off Route 1 east of the Hellisheiði heath, where the road starts to follow the river Hengladalsá. We drive on a lava field a short distance until arriving at a ford on the Hengladalsá, across from which is an intersection where we take a left turn and follow the river a bit upstream. The river is followed and crossed several times along this section. This is an ideal practice section for those who have seldom if ever driven across streams before. Here the Hengladalsá is a small river and there are clear breaks in the river which are followed. But if these are not followed, the driver quickly discovers how rapidly the stream deepens. The river also has many deep pools here and there in the calm stretches along the banks, and these are deep enough to damage vehicles if driven carelessly. Finally, we come to an intersection where the other track to Innstidalur is met. Tour operators often call this circuit *Þúsundvatnaleið* (The Thousand Lakes Track).

At the intersection, we continue up a rise, taking the left turn, where there is a gorgeous view across the valleys of Hengladalur and Fremstidalur. Inwards from the rise is the hut called Hreysi above the road. Along this section it becomes rougher and tougher going and therefore often requires shifting into low gear. When arriving at the next intersection, the tough road goes to the left to the hut. Above this hut, it is a short way to a pleasant area of hot springs and steam vents, beneath Hengill.

Sources exist that tell of outlaws that hid in Innstidalur, in the south-

On the way to Innstidalur. Photo: Jón G. Snæland.

ern part of Hengill, and evidence of habitation has been found up in the mountains.

From the previously mentioned intersection it is possible to drive farther up the valley and the route ends at a gap called Sleggjubeinaskarð.

Fields of flowers in Innstidalur. Photo: Jón G. Snæland.

Jökulháls

Start – Finish	**On Route 574 at Stapafell – On Route 574, east of the village of Ólafsvík.**
Distance/Time	**19 km. One hour's drive.**
Fuel	**At the village of Ólafsvík and at Arnarstapi on the peninsula's south shore.**
Caution needed due to	**No risks along the way.**
Passable for	**All vehicles, during summer.**
Of interest	**Snæfellsjökull glacier, vast panorama views, caves and shore formations at Arnarstapi.**

Jökulhálsleið is partially within the Snæfellsjökull National Park, which was established June 28, 2001. The route traverses the peninsula east of the glacier Snæfellsjökull. It is passable by all vehicles once the snow is gone and after it has been graded, but snow sometimes persists as late as July.

The Snæfellsjökull glacier is 1446 meters above sea level. On the edge of the crater at the top there are three small peaks, or prominences, known as Jökulþúfur. It is long since Snæfellsjökull erupted, but the mountain is known to have erupted at least 20 times. Three large ash eruptions have taken place: 9000, 3900 and finally 1750 years ago. Photo: Jón G. Snæland.

We start our trip up a steep slope (F570) near the mountain Stapafell, and ascend to a point near the mountain known as Heljarkinn. There is a parking place where travellers can pause and take a short walk to the cave called Sönghellir (Song Cave), so named for its acoustics.

From Sönghellir drive up the slope ahead where there is some timber debris along the roadside. This timber is what is left of old tracks used to slide carloads of cinder down the mountain. This effort began in 1937 and the cinder was broken into gravel and sent either to Reykjavík or directly exported. Above and to the east of the road are relics of old huts used by those working the cinder, but these have not been used since 1941.

There is a ski lift on Jökulháls and also a base camp where the company Snjófell has rented snowmobiles to visitors and offered snowcat trips to the summit of Snæfellsjökull. Snjófell also operates a restaurant, a guesthouse and campground in the community of Arnarstapi. At the highest point on Jökulháls there is a magnificent view across the bay of Faxaflói and along the peninsula of Snæfellsnes.

After crossing the highest point, the road begins to descend towards Breiðafjörður fjord, and the town of Ólafsvík. The drive is down a steep slope past Sandkúlur and Geldingafell and to an intersection. We keep straight on and eventually come to a small hydro-electric power plant, named Rjúkandi, for the waterfall of the same name in the Fossá river. The mountain road ends at a Roads Administration site on Route 574.

Now it is a good idea to go around the end of the peninsula, counter-

Snæfellsjökull glacier viewed from the east on the Jökulháls route. Photo: Jón G. Snæland.

clockwise along Route 574 and visit the seaside villages and see more of the National Park. Three villages dot the northwest shore of the peninsula: Ólafsvík, Rif and Hellissandur. They are now united under a joint munici-

At the Öndverðarnesviti lighthouse, just north of Skálasnagaviti lighthouse. There was a substantial settlement at Öndverðarnes in the past and ruins of houses are found on the peninsula. Photo: Jón G. Snæland.

pal government as the rural district of Snæfellsbær. Both Ólafsvík and Rif have harbours and have been significant fishing ports for many centuries.

At the southern boundary of the village of Hellissandur, we enter again the Snæfellsjökull National Park, but just before entering we pass a cluster of buildings called Gufuskálar. This used to be a station where some of the NATO forces had a LORAN navigation station, and the antenna here was once Europe's highest man-made structure, over 400 metres high. Now there is a rescue academy operated at Gufuskáli.

Outward from Gufuskáli is a track out to a small point, Öndverðarnes, where there are two lighthouses that are called Öndverðarnesviti and Skálasnagaviti, and there are also ruins of earlier settlements. Along Route 574 there is one natural gem after another. Among them are Saxhólar, Hólahólar and Djúpalón, and at the latter site are the remains of an old shipwreck. Then there are the spires of Lóndrangar and the quaint villages of Hellnar and Arnarstapi, which have fascinating shore formations and thriving bird life, to name but a few.

Kaldidalur

Start – Finish	**Intersection of Routes F550 and 52 – Route 518.**
Distance/Time	**40 km. One and a half hours.**
Fuel	**At Húsafell recreational site and in suburban Mosfellsbær.**
Caution needed due to	**No particular hazards.**
Passable for	**All vehicles.**
Of interest	**Glaciers and mountains.**

Kaldadalsleið lies south of the Bláskógaheiði heath and north to Húsafell, going between the tiny glacial vestige Ok (elev. 1170 m) to the west, and the larger Þórisjökull glacier (elev. 1329 m) to the east. Earliest mention of travels along Kaldadalsleið date from the 15th-16th century.

We start by taking the first left northward after descending to the lower level of Þingvellir National Park, which becomes Route 52. At the intersection which occurs after 23 kilometres, continue north straight through on the origin of Route F550 where the left turn goes west as Route 52 to Uxahryggir. Shortly to the east is Skjaldbreiðarvegur mentioned elsewhere in this book (pg. 139), but we continue north. Just above the intersection is

In his book Hálendið (The Highlands), Guðmundur Páll Ólafsson describes Prestahnúkur as follows: "Mt. Prestahnjúkur is an unusual rhyolite mountain and its colour alters according to the light because the rock is glassy. The height of the sun and amount of cloud affect its colour more than that of many other mountains. Rhyolite is called "Pearl-stone" because as it erodes it forms little balls or pearls. This type of rock is the result of very sudden cooling and solidifying of the lava. This makes Prestahnjúkur an unusual mountain in the highlands of Iceland." Photo: Steingrímur Steinþórsson.

an orange hut – an emergency shelter provided by Landsbjörg, a rescue association that operates many such huts throughout Iceland. Just north of the hut are relics of a weather and guard station operated in World War II to protect navy ship convoy traffic in and around Hvalfjörður fjord. This was a remote station and winter conditions must have been awful, but Icelanders provided supplies by horseback, including food, fuel and beer.

North of the hut the road wanders down and up slopes. On the right off the road is a famous old cairn of Kaldidalur, which locals call a *beinakerling* (Bone Crone) because such old cairns usually contain a shank bone of a sheep or cow, into which messages or rhymes for following travellers were placed. We also pass on the right a rugged track leading to the base of Þórisjökull glacier and later to the mountain Prestahnúkur, a colourful rhyolite mountain.

Farther north on Kaldadalsleið is a section called Sláttulág, which goes through an area known as Skúlaskeið. Skúlaskeið was such a rocky track that a man of the gentry, Bjarni Thorarensen, had it ploughed at his own expense in 1822.

Although the road is far from smooth even now, imagine what the first auto trip must have been like in 1928, when a couple of Icelanders drove a new Whippet Overland 1928 model along here on a journey that took them 20 hours! Even after improvements for the millennium celebration of the Alþing Parliament in 1930, it was still an arduous journey. North of the mountain Hádegisfell the road splits, with Kaldadalsleið going

Towards the north of Langjökull glacier, opposite Eiríksjökull glacier in an area called Flosaskarð pass, there is a large ice cave. Photo: Jón G. Snæland.

left. The right branch goes over a bridge on the stream Geitá, and in towards Langjökull glacier, where the hut called Jaki stands. Approaching Langjökull from Jaki is popular, as it is not very steep at this point.

On the road through Kaldidalur, the Geitlandsjökull glacial tongue in the distance. Photo: Óskar Hafþórsson.

From the ice cave at Flosaskarð. Photo: Jón G. Snæland.

From Kaldadalsleið, at this point, there is a view across the Geitlandshraun lava and inland to Mt. Hafrafell and to the glacier Eiríksjökull (elev. 1675 m). Kaldidalsvegur now continues down to Borgarfjörður in steep switchbacks along the Geitá river and the route ends at the junction with Route 518. From this junction it is a short distance south to the service centre at the recreational area of Húsafellsskógur, where various services are available to the traveller.

Kjalvegur

Start – Finish	**At the Visitors' Centre at Gullfoss waterfall – At Blöndustöð.**
Distance/Time	**173 km. Six to eight hours' drive.**
Fuel	**At Geysir, at Hveravellir in the highlands, and the town of Blönduós.**
Caution needed due to	**No particular hazards.**
Passable for	**All vehicles during summers.**
Of interest	**Glaciers, hot springs, natural thermal pools and waterfalls.**

The start of our trip is the Visitors' Centre at Gullfoss waterfall, undoubtedly the most famous waterfall in Iceland. It is on the Hvítá river and falls 32 metres in two main drops into the canyon Hvítárgljúfur.

After seeing Gullfoss, we head north for the highlands along Kjalvegur (Route 35). For a few kilometres beyond Gullfoss, Kjalvegur is a good, paved road, but after that it becomes a coarse gravel road that is often riddled with potholes, 'washboard' ripples or huge puddles in wet weather.

The hot pool at Hveravellir was built around 1950 and is extremely popular. Ten years later, around 1960, the pool and the whole hot spring area was declared a protected site. According to the F.Í. Annual the pool was completed or rebuilt in its present form in 1985. Photo: Þóra Sigurbjörnsdóttir.

Some kilometres after the gravel starts, we arrive at a glacial river, the Sandá, where there is an open Quonset hut used by shepherds and other travellers. Furnishings inside are meagre, nevertheless it has been given the comic name 'Hotel Sandbúðir'. From the 'hotel' we continue onwards, and there are several intersections or forks in the road, the most significant of which will be described as we meet them. We cross the Grjótá river and north of it is a hiking trail up into an area known as Kórinn, a unique dell at the Valagil canyon, in the west part of the Bláfellsháls rise. There is a cave at Kórinn that was used as a shepherds' shelter in the old days.

On Bláfellsháls is a track that leads west to Skálpanes, where tour operators offer snowmobile trips onto Langjökull glacier. There is a large cairn at the roadside on the rise and travellers often stop to add a stone. The story goes that this cairn was started in 1933 when a road worker left his street shoes by the roadside and put a few stones on top of them to help him find them later. Others came along and added stones until gradually the cairn we see today was formed.

Now we are nearing the lake Hvítárvatn and the river Hvítá that runs from it. Hvítárvatn is a large glacial lagoon into which fall two glacial tongues of the larger Langjökull. These are Suðurjökull and Norðurjökull and between them is the mountain Skriðufell. On the north bank of the lake is the oldest hut run by Ferðafélag Íslands, built in 1930. It is said to be haunted. A bit north of the Hvítá river on the roadside is a guest hut called Árbúðir, which is run by the travel agency Ferðaþjónusta Gljásteins.

A cairn built in memory of a famous mountain traveller named Hafþór Ferdinandsson (Hveravellir dasher) who visited Hveravellir often, summer and winter. Photo: Jón G. Snæland.

This service also sells accommodation in the hut called Gíslaskáli, south of the Svartárbotnar bottoms, and in Fremstaversskáli at the base of Bláfell. Past the area of Fremri-Skúti, two mountains come into view to the left/west. First is the palagonite mountain Baldheiði, formed like all palagonite under a glacier. North of it is a massive mountain called Hrútfell which is topped by a glacial cap. To the right/east are the Kerlingarfjöll mountains, and north of them, the glacier Hofsjökull.

Beyond Innri-Skúti is the road to Kerlingarfjöll (pg. 23, 82). Just north of that intersection is a track down to Gíslaskáli and to the north is the swell Fjórðungsalda, which has a viewpoint disk on top that is interesting to see. Visible to the north from Fjórðungsalda is Hofsjökull glacier, and a very difficult track leads there. Its start is clearly visible on Kjalvegur where there is a sign warning travellers of quicksand and rivers with stiff currents. But despite these obvious warnings, foolhardy travellers with inadequate vehicles have tried the route and got into difficulties. Therefore, we specifically re-emphasize the risks on this route.

Southwest of the mountain Dúfunefsfell, we arrive at a sign pointing towards Hveravellir. Those driving Kjalvegur should not miss the chance to visit Hveravellir. Various tourist services are offered at Hveravellir, and it is possible to purchase fuel and overnight accommodation here. There

The old turf hut at Hveravellir. It was built in 1922. Photo: Jón G. Snæland.

are two huts run by the group called Hveravallafélagið – a new hut built in 1980 and an older one built in 1938. Below the older hut is a turf hut that was built in 1922. It is believed to have been built at the site of the first highland shelter in Iceland, one described in the Sagas of Sturlungar, a collection of sagas written in the 12th and 13th centuries.

Iceland's most famous outlaws, Fjalla-Eyvindur and his partner Halla, also lived here for a while and it is possible to see their cave just south of the hot springs area. The thermal area here has many different and beautiful hot springs that have attracted visitors over the centuries.

When we return to the intersection and resume travel northward, we pass a small airfield, then the mountain Dúfunefsfell, and cross a bridge over the river Seyðisá, at which point we drive on a raised gravel road. North of Seyðisá is a small old shepherds' hut, which has recently had its stone walls renovated. This hut is called Kúlukvíslarkofi, and at the site called Áfangi is another hut where meals and overnight accommodation are sold.

The road continues to Blöndulón, a reservoir behind the hydro-electric dam Blönduvirkjun, which is a little further to the north. Guided tours of the power plant are available for visitors. Our route continues north ending in the Blöndudalur valley.

Kvíslaveituvegur

Start – Finish	**Mt. Vatnsfell – The dam on the Þjórsá river.**
Distance/Time	**76 km. At least three hours.**
Fuel	**At the Hálendismiðstöð station at Hrauneyjar.**
Caution needed due to	**Driving across high power dam.**
Passable for	**All jeeps and vehicles with high clearance.**
Of interest	**Ancient settlement sites, beautiful glaciers and waterfalls.**

The approach to the start of this route takes us near the hydro-electric power plant called Vatnsfellsvirkjun, where there is a sign labelled 'Sprengisandur'. Here we drive below a large dam and up a steep slope onto Vatnsfell mountain, where we are on a raised gravel road, which we follow to the Þjórsárstífla dam. The road is very rough and in many places large, immovable rocks punctuate the surface and can puncture tyres, so it pays to drive carefully. This part of the road was built by Landsvirkjun National Power Company in 1980-1984, in conjunction with construction of the diversion dam at Kvíslaveitur.

From Vatnsfell we drive alongside a large reservoir that is named Þórisvatn, and later down into a valley at the base of the swell Ósalda. There is a sign which is lettered with 'Fagrifoss'. Fagrifoss (Pretty Falls) is a waterfall that no traveller should miss in the sizeable glacial river Kaldakvísl, the current of which is sometimes diverted into the reservoir at Sauðafell mountain. If the current is diverted, the reduced water in Fagrifoss is clear, otherwise it is clouded with glacial sediment. Either way, the waterfall is always quite pretty.

There is a great dam by Mt. Vatnsfell and it is exciting to drive down to it a short way from the Sprengisandsleið to see the power of the water close up. Photo: Jón G. Snæland.

Driving on the glacial tongue Þjórsárjökull which comes out of Hofsjökull glacier. The mountain in the centre of the picture is called Arnarfell the great and to the right of it is the glacial tongue Múlajökull, also from Hofsjökull. Photo: Jón G. Snæland.

The waterfall Fagrifoss in the river Kaldakvísl just off the Sprengisandsleið route. Photo: Jón G. Snæland.

Just this side of the track to the falls, we drive over a bridge spanning a large canyon of the Kaldakvísl currents, which is worth seeing. To the north we arrive at an intersection where Sprengisandsleið is to the right, but we continue straight on Kvíslaveituvegur. At a point even with Kvíslarvatn lake, it is barely three kilometres to a track leading down to Þúfuvötn lake and the headland on it called Sóleyjarhöfði. At Þúfuvötn is a hut that goes by the name of *Páfagarður* (The Vatican). It is possible to drive a difficult seven-kilometre track to Sóleyjarhöfði, which is a perfect place to view the river Þjórsá, Iceland's largest river. There is also a great view to Hofsjökull glacier and the marshes of Þjórsá. These are collectively called Þjórsárver, and comprise the largest wetland oasis in the central highlands, as well as being the world's largest nesting area of pink-footed geese. The Þjórsárver were officially protected in 1981.

Beneath Sóleyjarhöfði is an ancient ford still used today by Icelandic jeep drivers. In the ford there is a large island that is driven to, and from there the direction is towards an old turf house on the west bank of the river. Those not familiar and experienced with the ford are strongly advised to avoid it.

The huts at Gásagustur, just west of the Kvíslaveituvegur road. Mt. Arnarfell the great in the distance and to the left the Múlajökull glacial tongue creeps down from the Hofsjökull glacier. Photo: Jón G. Snæland.

North of Þúfuvötn lake is a hut called Gásagustur that draws its name from the fact that just north of the hut in the Þúfuver marsh was the starting point for trips on the old Sprengisandsleið, when the sands were called Gásasandur. In the old days people frequently travelled across Sprengisandur, and, for example, the Bishop regularly went north on church business over Sprengisandur. However, during the years 1740–1741 it seems travel across the 'Sands' stopped for about 30 years. Fjalla-Eyvindur (Eyvindur Jónsson, born 1714) and his partner Halla Jónsdóttir lived together as outlaws in the wilderness for at least 15 years, including a continuous interval from 1765 to 1772 at Þjórsárver marsh, and built five huts at the stream Eyvindarkvísl. The author, Guðmundur Guðni Guðmundsson, writes that Fjalla-Eyvindur spent a total of 40 years as an outlaw in the wilderness.

In 1772, the Parliamentary Secretary Sigurður Sigurðsson asked Einar Brynjólfsson to travel Sprengisandsleið, and Einar set out with four other men on August 3, 1772. Einar's first choice as a guide, Eiríkur Hafliðason, who earlier had recorded the details of the route, declined the trip due to advanced age. Einar Brynjólfsson and his group went first to Hald, at the southernmost point of Búðarháls, and crossed the Tungnaá at a ford. When they had gone north of Biskupsþúfa, they encountered Fjalla-Eyvindur and Halla, captured them and transferred them north to Jón Tómasson, a district administrative officer. Eyvindur soon escaped from his cell there and went south again to Eyvindarver and took various stuff from his huts. He was gone by the time Einar and his group returned south.

Two years later, Eyvindur released Halla from her imprisonment. If Eiríkur Hafliðason had been travelling with Einar, it is not unlikely they would have gone along Arnarfellsvegur if the Þjórsá had had a lot of water in it, and Eyvindur and Halla would not have been captured. A posted track leads to Eyvindarver and there is also a short, marked track to Hreysi. East of Eyvindarver is a good road that connects Kvíslaveituvegur with Sandölduleið. North of Þúfuver we drive alongside several dams and finally arrive at Þjórsárlón in the wetlands Háumýri where we cross a bridge over a stream leading out of the Þjórsárlón reservoir.

Lakaleið

Start – Finish	**Route 1 at Hunkubakkar – Laki craters.**
Distance/Time	**48.8 km to Laki, loop drive, 24 km. 3-4 hours.**
Fuel	**In the town of Kirkjubæjarklaustur.**
Caution needed due to	**Stream fords.**
Passable for	**All jeeps during summers.**
Of interest	**Waterfalls, volcanic craters, lavas and abandoned farms.**

Turn north off Route 1 and onto Route 206. Soon we cross a prominent glacial river, called Skaftá, which has its source in the giant glacier Vatnajökull. After crossing the river we climb a slope passing a few farm houses and come to an intersection at the start of the mountain road leading into Lakagígar. If we drive a bit further past the intersection there is a place to park the car and walk to the right and view the beautiful river canyon Fjaðrárgljúfur, which is rugged and deep.

The Lakaleið is easily passable, but somewhat slow-going with many slopes and bends in the road, as well as numerous brooks to cross.

Eventually, we come to an intersection with a line road that follows a high tension line. But we continue straight and just north of the intersection to an abandoned farm known as Eintúnaháls. This is an ideal place

The abandoned farm at Eintúnaháls which was occupied up to 1934. Photo: Jón G. Snæland.

The waterfall Fagrifoss in the river Geirlandsá. On the way north to Laki the route fords Geirlandsá. There you can see a track to the right that leads to Fagrifoss. Photo: Jón G. Snæland.

to stop for a bite to eat. Eintúnaháls was abandoned in 1934 after more than a century of farming dating back to 1826.

The first river ford of any consequence is across the river Geirlandsá, which has the aptly named waterfall Fagrifoss (Pretty Falls) in it. Cross the river and to the right is a track about 300 metres long to the waterfall. Travellers should not miss a stop at this incredibly beautiful waterfall. After the waterfall, we continue north and cross a few small brooks and arrive at a ford across the river Hellisá, which is the largest stream on our journey. The ford is reliable and the flow is light at the ford.

A bit farther north Routes F207 and F206 intersect. It is possible to choose either direction in to Laki, where the journey ends. These roads form a ring, but it should be emphasized that F207 is a rough jeep road. Off of F207 is a road down to the hut in Blágil where the nature warden for the Laki area is based. It is possible to book overnight accommodation here using your own sleeping bag.

At Blágil we are in the midst of a volcanic area formed by volcanic debris from the craters and fissures of the eruption known as the Skaftáreldar (1783–1784). From Laki there is a great view toward Vatnajökull and along the row of craters collectively called Lakagígar, or Eldborgarraðir. There are also posted hiking trails through the area. The crater row is nearly 25 kilometres long and includes 135 craters. The Skaftáreldar eruption was one of Iceland's largest eruptions, producing the third-largest amount of lava that has erupted since the end of the Ice Age. The eruption started with tremendous ashfall and a strong smell of sulfuric acid on Whitsun (Pentecost Sunday) June 8, 1783. Lakagígar lies on 10 parallel fissures, each being 2-5

Fjaðrárgljúfur canyon is an awesome canyon around 100 meters in depth just off the route to Lakagígar. Photo: Jón G. Snæland.

km long. Statistics tell the story of just how powerful this eruption was, and why its impact was so far-reaching. The volume of the lava flow is 15 cubic kilometres, of which 14.5 cubic kilometres are actual lava, and the rest being loose debris such as ash, cinder and bombs. The eruption belched forth 400-500 million tons of gases. The lava fountain is said to have reached a height of 800-1400 metres, and the eruption cloud soared to heights up to 15,000 metres (46,000 ft). The eruption brought tragedy and disaster to the country, and was called *Móðuharðindin* or "The Haze Hardship". Poison ash fell throughout Iceland. Livestock died of starvation and a haze polluted the air. One out of five Icelanders died as a result of the eruption.

Landmannalaugar (Sigölduleið)

Start – Finish	**Hrauneyjar – Landmannalaugar**
Distance/Time	**40.7 km. At least one hour.**
Fuel	**At the Hálendismiðstöð station at Hrauneyjar.**
Caution needed due to	**Small fords.**
Passable for	**All vehicles during summers.**
Of interest	**Colourful mountains and natural thermal pools.**

From the Hálendismiðstöð Highland Centre at Hrauneyjar we go a short distance north on Sprengisandsvegur, Route F26. This option to Landmannalaugar skirts the reservoir Hrauneyjalón and crosses a thrilling bridge over a manmade diversion channel below the station-house of the Sigölduvirkjun hydro-electric power plant. While crossing the bridge, it is worthwhile checking to see if there is a waterfall in the channel. When the Krókslón reservoir above the dam is emptied, a powerful waterfall, called Sigöldufoss, is created. It only occurs when the reservoir is emptied or work is being done on the dam. Across the bridge is a pretty view across Krókslón and north of the glacial river Tungnaá. We drive along the lava and the Tungnaá, which is one of Iceland's largest rivers.

Just before reaching a cattle grid in the road, there is a faint track down to the river. Here it could be fun to stop and view the tremendous current of the river before following the track to the river to see the old boat house

The landscape at Svartikrókur is magnificent. The picture is taken at Svartikrókur over the Tungnaá river, with Skyggnisalda swell in the background. Photo: Jón G. Snæland.

The Sigöldufoss waterfall. When power plant workers empty the Krókslón lagoon is the water is channelled into a river bed, forming a powerful waterfall called Sigöldufoss. However, there is no water in it except when there is constuction by the dam, which may be at only 20 year intervals. Photo: Jón G. Snæland.

on the riverbank. This was a storage hut for the ferry boat that was used to ferry people and livestock across the river at a crossing point called Bjallavað. There on the other side is another boat hut.

At the mountain Eskihlíðarhnausar another faint track is visible leading west into the valley Dómadalur, which is not taken. Continue driving past the mountain. When past the mountain, we see optional tracks leading east towards the Tungnaá river, as well as a few posts. This is the way to the ford site Hófsvað, *which should be avoided*. It is only two kilometres

from the road, and used to be the site where the Tungnaá could be crossed by autos.

From the mountain, we now drive along sand flats and see two mountains ahead of us: Hnausar and Tjörvafell, and between them is a small crater lake, Hnausapollur, which has a loop approach, allowing vehicles to drive to the lake and back on to the road, without turning around. Hnausapollur has also been called Bláhylur, and, in ancient times, Litlavíti.

A little farther on is an intersection where it is fun to divert to the left to see Ljótipollur before going further. Ljótipollur, or Stóravíti as the crater was called in the old days, is a crater lake which was formed after an eruption in 1480. Ljótipollur means Ugly Puddle, and it could hardly be more misnamed; the water and crater are incredibly beautiful and colourful. With care it is possible to drive to the crater's rim.

From Ljótipollur, we return to the intersection and turn left to drive over the rise at Frostastaðaháls above Frostastaðavatn lake. Once over the rise, turn right and drive along the base of the mountainside of Suðurnáma. As Landmannalaugar is approached, we cross a small river called Námakvísl, and then a small brook, Laugalæna. These streams are generally easily crossed by jeeps, even smaller ones. Those who do not trust themselves to drive across the rivers can leave their vehicles and walk in to Landmannalaugar, using pedestrian bridges over the streams.

The boat hut at Bjallar ford. Photo: Jón G. Snæland.

The natural hot pool at Landmannalaugar is one of the most popular tourist destinations in Iceland. Photo: Þóra Sigurbjörnsdóttir.

Various services are offered at Landmannalaugar. Ferðafélag Íslands has a large guest hut there and sanitary facilities as well. There is also a large campground, and from Landmannalaugar to Þórsmörk is the ever-popular hiking trail Laugavegurinn. Hikers usually start at Landmannalaugar. Landmannalaugar also has Iceland's most popular natural thermal pool, which attracts tens of thousands of visitors annually.

Landmannaleið (Dómadalsleið)

Start – Finish	**Landvegur, 26, east of Búrfell – F208, by Frostastaðavatn.**
Distance/Time	**20.2 km. Just more than one hour.**
Fuel	**At Hrauneyjar and at Leirubakki.**
Caution needed due to	**One ford and several creeks.**
Passable for	**All vehicles with high clearance during summers.**
Of interest	**Beautiful mountains, caves, lavas and classic farms.**

Landmannaleið (F225) starts as a turn off Route 26 just east of Mt. Búrfell, going east. After driving across a sandy area there is an intersection at a place called Alda. The north track goes to Áfangagil, and a south track leads to Mt. Hekla, but we continue east, with Hekla dominating the landscape for a while.

It is not known whether Hekla had been climbed before the natural scientists Eggert Ólafsson and Bjarni Pálsson reached the summit on June 20, 1750. It is believed nobody dared attempt it because of the legend that Hekla was the gates to hell, where lost souls were damned to the eternal inferno. Hekla is an active volcano and erupts regularly, often causing

Hekla, one of Iceland's most famous volcanos, was in its time thought to be the gateway to Hell. It has erupted repeatedly in historical times. When it erupted in 2000 they were able for the first time to predict an eruption with real accuracy. At 18:00 on February 26th it was announced in a radio newscast that an eruption was imminent and the mountain erupted at 18:17. Photo: Jón Ebbi Halldórsson.

The cave Landmannahellir was for centuries used to house sheep and horses. Photo: Jón G. Snæland.

considerable damage. Hekla's first eruption after settlement was 1104, followed by eruptions in: 1158, 1206, 1222, 1300, 1341, 1389, 1510, 1597, 1636, 1693, 1766, 1845, 1947, 1970, 1980-81, 1991 and 2000.

With the eruption of 2000 it was for the first time possible to predict eruption with some accuracy. It was announced in a radio newscast at

18:00 on February 26th that an eruption of Hekla was imminent and 15 minutes later the eruption occurred.

Upon entering the protected area of Fjallabak, we arrive at an intersection where a jeep trail, usually called Krakatindsleið, to the right/south leads to the back side of Hekla, so to speak. But we proceed eastward. Just after this intersection we cross the harmless stream Helliskvísl, after which the route forks with our route Landmannaleið going right across another stream, Rauðufossakvísl, at a ford. However, the left branch goes to Landmannahellir, crossing Helliskvísl again, and circling the mountain Kringlan.

If the route to Landmannahellir is taken first, we cross Helliskvísl as previously mentioned and continue along it until a large cluster of huts is reached. Here a company called Hellismenn ehf. offer various services June 19 – September 5. Among options include permits to camp and fishing permits for nearby lakes. The oldest standing building here is an old shelter built with public money in 1907. Initially there was a sleeping loft for several persons, but this is ruined and now the place is usually used for emergency shelter for tent campers in bad weather. There is still a table with benches here. Just south of the hut is a cave that was long used for horses or for shelter for the shepherds of the Landmannaafréttur high pastures. It could hold 70 horses, while men stayed in a shallow cave nearby with a rock wall at the cave mouth.

When leaving the area of Landmannahellir, we again cross Helliskvísl near Lóðmundarvatn lake and meet up with Landmannaleið, just south of a small river known as Klukkugilskvísl, and we go left/east. We then cross the rise at Dómadalsháls and descend into the valley Dómadalur and cross the Dómadalshraun lava to the intersection of Landmannaleið and Fjallabaksleið, Route F208.

Langavatnsdalur

Start – Finish	**Svignaskarð gap in Borgarfjörður – Hörðudalur in Dalasýsla District.**
Distance/Time	**38.8 km. Three to four hours' drive.**
Fuel	**In the towns of Búðardalur and Borgarnes.**
Caution needed due to	**Fords, steep slopes and muddy spots in wet weather.**
Passable for	**Slightly modified jeeps, but not solo.**
Of interest	**Mountains, beautiful lakes and waterfalls.**

Langavatnsleið is an ancient cross-country route between Borgarfjörður fjord and the district of Dalasýsla and it is mentioned in the Icelandic Sagas. From the south we leave Route 1 left/northeastward at a site known as Svignaskarð about 17 km north of the town of Borgarnes in Borgarfjörður. A sign here is labelled 'Langavatn'. The route passes through a popular area for summer cottages along the river Gljúfurá, which courses in canyons through beautiful shrub-land of birch and willows.

As we approach the rise at Beilárháls at an elevation of 340 metres, we cross the river Gljúfurá on a bridge and go up a slope to the top of a rise. At the top of the rise, we are greeted with a beautiful view over Langavatn lake

Driving across a shallow ford in the Langavatnsá river. Beyond is the Langavatnsmúli ridge, along which there is a track leading via Grenjadalur valley to built up areas. Photo: Jón G. Snæland.

and Réttarmúli, as well as Langavatnsmúli. Langavatn is at an elevation of 215 metres above sea level, and is 5.1 square kilometres in size and 36 metres deep at its deepest. We descend the north side of the rise down to Beilárvellir fields, as the flatlands are called at the base of Réttarmúli. It is here that the river Beilá enters Langavatn lake.

The river Beilá is crossed by an old bridge, which has an impressive iron frame whose overhead beams are too low for high vehicles. Such vehicles have to drive a bit downstream from the bridge and cross the river via a ford. The water volume in the Beilá is minor, and hence it is rather obvious where the river can be forded at a long, pretty and flat stretch below the bridge. Just north of the bridge is a shepherds' hut where it is possible to purchase overnight accommodation.

Once we have travelled farther north of the hut, the route becomes a bit more difficult. It gets rocky and many small brooks traverse the track. Once we are beyond Langavatn lake, the river Langavatnsá must be crossed. The ford is shallow, but travellers must examine the ford well before entering the river and proceeding into the flatlands and bottom of the Langavatnsdalur valley.

Having crossed the Langavatnsá, we drive along a flatland stretch of dirt road, which can be treacherous in wet weather. We continue to the north part of the valley, and an intersection appears at the north slopes of the valley. The track to the left is not taken as it leads via a rough road back to settlement along Langavatnsmúli and down Grenjadalur valley.

A herd of horses in Laugardalur valley. Icelandic horses are often allowed to run free in the highlands over the summer and it is not unusual to run across a large herd when off the beaten track. Photo: Jón G. Snæland.

We on the other hand, follow the north slopes of the valley past the valley of Fossdalur, where there is a beautiful waterfall that nobody should miss. In front of the valley Víðidalur, we turn west and go up a steep slope, up the valley and drive here on the edge of a canyon, then across a river and up another steep slope. Then the valley is followed up to an elevation of

In her book *Söguferð bæ frá bæ*, which is an historical journey from farm to farm in the area, Ingibjörg Bergþórsdóttir tells of two tragic events that occurred in the old days when there was a sizeable settlement in Langavatnsdalur:

"... in the year 1811 a man named Sæmundur moved here with his family and built a farm on the ruins of the old farmstead Borg. In February of 1813 the fire died in the cottage and the farmer froze to death at Staðartunga when he was trying to get new fire. The wife and children continued to live in the cottage and barely survived the winter months. One of the children went to the settlement seeking help without much success and stole a horse on the way back to the cottage, where the horse was slaughtered for food. When convicted of horse theft, the family was severely punished."

Torfhvalaskáli hut on the east side of Langavatn lake at the foot of Réttarmúli promontory. Photo: Jón G. Snæland.

about 400 metres at Sópandaskarð pass. After the highest point, the road then descends steep slopes into the valley of Laugardalur. These slopes have loose gravel and can prove difficult for smaller jeeps, especially if the road is driven in the opposite direction uphill to the south. The Laugardalur valley leads down to another valley, Hörðudalur, along beautiful narrow passages and fields between high mountains. Use caution as there are often horses that roam free here.

When we emerge from Laugardalur to the north and Hörðudalur valley opens, we arrive at the ruins of an old building and pool, which used to be the community pool for residents of Hörðudalur. As our journey ends, above the road about 20 metres away there is a small, natural thermal pool rimmed with stones, whose stonework has started to deteriorate.

Loðmundarfjörður and neighbouring inlets

Start – Finish	**Hvannstóðsdalur in Borgarfjörður – Úlfsstaðir, Loðmundarfjörður.**
Distance/Time	**32 km, from Borgarfjörður to Loðmundarfjörður. About an hour, with no stops.**
Fuel	**Bakkagerði at Borgarfjörður eystri.**
Caution needed due to	**Small rivers and steep slopes.**
Passable for	**All jeeps during summers.**
Of interest	**Abandoned farms, colourful mountains and unique nature.**

The mountain road over Loðmundarfjörður starts innermost at Borgarfjörður eystri on Route 946 at the mouth of the Hvannstóðsdalur. Here the track threads its way up into the mountains ahead, entering inlets and abandoned valleys. Once arriving above the site Afrétt, and at the base of the mountain Hvítserkur, a track is visible to the north that leads down to Breiðavík, which will be discussed later in this section.

We now drive between the mountains of Hvítserkur and Náttmálafjall along the Húsavíkurheiði heath. It is a bit steep here, but passable for all

The East Iceland Travel Association's hut, Húsavík, in the inlet of that name. Photo: Jón G. Snæland.

four-wheel drive vehicles. The highest part of this route has an elevation of 486 metres above sea level, at Mt. Hvítserkur, which is sometimes also called Röndólfur. The latter name derives from the unique colour pattern in the mountain. Hvítserkur is made of white and pink-coloured rhyolite, with black basalt veins crossing through it forming the Roman numeral XI.

Steep slopes take over now down to Húsavík, but higher in the inlet there is an intersection. One route leads to Húsavík, a hamlet with its own church, Húsavíkurkirkja, but the adjacent farm was abandoned in 1974. All other farms here have been abandoned. These, with the year they were abandoned, are: Dalland in 1954; Dallandspartur in 1948; and Hólshús in 1943. Up the valley, there is a bridge over a stiffly flowing river, the Víkurá, and across it, the touring club, Ferðafélag Fljótsdalshéraðs runs a nice guest hut called Húsavíkurskáli.

Now we proceed up the steep-sloped Húsavíkurheiði heath along Nesháls rise and over into Loðmundarfjörður. Along the way the track is 451 metres above sea level at its highest, and consequently there are often snow drifts here well into the summer. Over the rise, we drive down the mountain Skælingur, which sailors often call the 'Chinese Temple' because from the sea Skælingur resembles a Chinese Temple. The slope down into Loðmundarfjörður is very steep, but the road is good as it winds all the way down to sea level. Some old ruins are visible from the road. We have to drive across a brook at a ford and across the bridge over the river Stóra-Hrauná, where there is a pretty waterfall just above the bridge.

The bird life in Borgarfjörður eystri is lively. Here two puffins have settled on a shelf in the cliff. Photo: Kjartan Gunnsteinsson.

A farm on the north side of the valley is called Stakkahlíð, which is part of the Icelandic Farm Holidays system, and the house has an interesting collection of stones along one wall. There is no longer any permanent settlement in Loðmundarfjörður, where the pioneer settler was Loðmundur the Old. He lived three winters here before moving south to the sands of Sólheimasandur where his high seat posts washed ashore. This is the same technique used by the viking Ingólfur Arnarson who first settled Reykjavík. In its hayday, between 1800–1900, 87 people lived in Loðmundarfjörður, but gradually residents dwindled and during the years 1940–1965, five farms were abandoned, the last one being Sævarendi in 1973.

There is a known mineral mine in Hraunsdalur valley that can be reached by a difficult jeep trail. Higher up the valley, a new hut has been built by Ferðafélag Fljótsdalshéraðs near ruins of the church site known as Klyppsstaður, where a church, dating from 1891, still stands. It is possible to drive farther into Bárðarstaðadalur valley, and to do so, we have to drive over the Norðurdalsá river, which is rather sizeable and has a stiff current. Beyond the ford at Úlfsstaðir is a house that is being renovated. Beyond Úlfsstaðir are the ruins of Bárðarstaðir, and it is impossible to drive further into Bárðarstaðadalur, but there is an old riding trail leading inwards along the valley and up to the Hérað district.

Mt. Hvítserkur (Röndólfur) is a particulaly beautiful and colourful ignimbrite mountain. Photo: Jón G. Snæland.

Here we turn around and return along Loðmundarfjörður, and Húsavík, and past Hvítserkur, until we come to a lane leading into Víknaheiði heath. It is a slow track that leads along the heath and down into Breiðavík via an unbelievably colourful chasm. There is another guest hut in Breiðavík, run by Ferðafélag Fljótsdalshéraðs, that is just as well-equipped and hospitable as others run by the club in the area.

There are a great many torrential rivers in the Víkur. Photo: Jón G. Snæland.

The majestic Dyrfjöll (Doorway) mountains are the highest in the fjord Borgarfjörður eystri. They are named for the huge opening (856 meters above sea level) between peaks either side, the highest being 1136 meters above sea level.

The Stórá river runs to the sea near the hut and it is crossed three times at fairly recognizable fords with reliable bottoms. The northernmost channel is the deepest, and the crossing is shallower if it is crossed to the right in the ford. The old namesake farmsite, called Breiðavík, stood only a few metres from the hut and was abandoned in 1943. Another farm, farther into the inlet, Litlavík, was abandoned in 1945 and ruins of it are visible to the south in the inlet.

Out of Breiðavík we drive up to Gagnheiði heath, where the track leads along a steep scree slope of rhyolite, which is called Hestur. Here the track ascends in switchbacks whish presents a real challenge to those in the smallest jeeps,. Over a short section the road climbs to an elevation of 479 metres and there some drivers find themselves suffering vertigo.

At the top of Gagnheiði is a fantastic view, among other things, over to Brúnavík inlet, which will be mentioned later. Our route continues down a ridge known as Þrándarhryggur, where in the middle of the ridge there is a unique stone that attracts attention, as on it is written, "A man is on the middle of Þrándshryggur, says little – but half-way to the inlet – Jón Sig."

Next, we drive along the base of an imposing mountain called Svartfell, and then down to the lowlands on Route 947. Once we arrive there, it is possible to drive out along the fjords, where soon a track entering the mountains is seen. This is a slow track over to Brúnavík. It leads up to the pass Hofstrandarskarð between the mountains Svartfell and Geitfell, and a short way in the valley down to Brúnavík, where it is necessary to walk out to the inlet itself, if that is a goal. On the shore there is an old emergency hut.

When our mountain trip is finished, it is not a bad idea to visit the village of Bakkagerði at Borgarfjörður eystri, to see the harbour there and the very diverse birdlife, among other things.

Mælifellssandur

Start – Finish	**Hvanngil canyon – Hólmsá river.**
Distance/Time	**22.5 km. About an hour.**
Fuel	**At the town of Hvolsvöllur.**
Caution needed due to	**Possible sand storms and wet, sandy conditions.**
Passable for	**All jeeps during summers.**
Of interest	**Glaciers, sand barrens and colourful, beautiful mountains.**

We start this route by driving from Hvanngil canyon to Syðri-Fjallabaksleið and driving south to the stream Kaldaklofskvísl, which is just south of Hvanngil. Kaldaklofskvísl is mixed with glacial water and can therefore change volume quickly, either in hot summer afternoons, when the glacier melts more, or in wet weather. It is fairly clear how the river is forded, because large stones have been placed in an arc across the river downstream from the ford, to keep drivers from going too far downstream. Other boulders can be carried by the river into the ford and cause problems, and there is also a slight sand ridge in the middle of the river that can be a problem for smaller vehicles. Once the river is forded, we turn immediately to the left and on to Route F210, which is the bona fide start of the highland part.

Some ugly scars left by off-road driving, which is forbidden in Iceland. Photo: Jón G. Snæland.

Initially, the track is very vague for a section of several hundred metres of rock slabs just south of the intersection. But it pays to follow the canyon of the stream Kaldaklofskvísl until the track becomes obvious on the other side of the slabs. We are now in the area known as Mælifellssandur, but in the old days the west part was called Gásasandur. This sandy area makes for quite comfortable driving and the view is spectacular, with the glacier Mýrdalsjökull only two kilometres away to the south, at its closest point to the road. It is not uncommon for super-jeeps to be driven up onto Mýrdalsjökull during winter. On older maps this part of the glacier was called Goðalandsjökull. Along this section, and all the way to the river Hólmsá, there is a risk of sandstorms, and therefore it is not recommended to visit during dry or windy weather.

Soon after entering Mælifellssandur, a small sand dune is visible on the left, which is named Slysaalda (Casualty Swell), so-called because four people from the Skaftafell area froze to death here in 1868. Their bodies were not discovered until 10 years later and their belongings remained there until well into the 20th century. A hundred years later, relatives put up a memorial to the victims at Slysaalda. Unpredictable winter weather in Mælifellssandur has claimed the lives of other Icelanders. For example, two men froze to death when they tried to walk to an inhabited area after their helicopter crashed during the winter of 1977, and in the winter of 2010 three travellers in an SUV got caught in bad weather, with the result that two, a man and a woman, perished near the ford at Kaldaklofskvísl when trying to walk out. Their companion, a woman, survived after being

Travellers by the monument on the Slysaalda swell. Photo: Jón G. Snæland.

found by rescuers. The message here is to avoid the circumstances in the first place, and if you become stranded, never, ever leave your vehicle, since it is more easily found than an individual in the wilderness.

Up on the sands, we head for Mt. Mælifell which rises majestically above them. But shortly before reaching Mælifell there is an optional fork in the road to the left, that runs between Veðurháls rise and Mt. Mælifell, leading to the hut called Strútsskáli which is owned by the touring club Ferðafélagið Útivist. From the hut it is a short walk to the well-known natural thermal pool called Strútslaug.

However, we do not go to Strútsskáli, but continue straight and cross a stream with the funny name Brennivínskvísl (Schnapps Branch). At this point we are at the southern base of Mt. Mælifell and our route bends east of the mountain, arriving at an area of several brooks. These are small and sometimes wander about the sands for quite a distance, and therefore it is best to watch out for wet sands where vehicles can become stuck. Across the brooks is an intersection, where the right option leads to F232, the Öldufellsleið that is mentioned in another section of this book (pg. 172). If we continue straight along F233, the route is the Álftakróksleið, on which it is possible to drive and join the Fjallabaksleið nyrðri, but along the way

A foot bridge crossing the river Kaldaklofskvísl by Þrætutangi. Photo: Jón G. Snæland.

are substantial bodies of water to cross. The first of these is the Hólmsá, which is a short ways from the above-mentioned intersection, and another river to cross is the Syðri-Ófæra.

Skagafjarðarleið

Start – Finish	**Gil in Vesturdalur valley – Laugafell north of Hofsjökull glacier.**
Distance/Time	**59.3 km. About two hours.**
Fuel	**At Varmahlíð and in the town of Akureyri.**
Caution needed due to	**Ford of glacial river.**
Passable for	**All jeeps.**
Of interest	**Natural thermal pools, uninhabited wilderness and glacier view.**

We start by driving Route 752 south from Skagafjörður fjord and enter the Vesturdalur valley. Below the farm called Gil, the actual mountain route starts into the highlands, with a fording of the river Giljá. The river is not large and generally has a good bottom with small gravel on it. The route skirts the east side of Vesturdalur, southward to Þorljótsstaðir, which is an old abandoned farm where a summer cottage has been built.

At Þorljótsstaðir the road ascends zig-zagging up 260 metres to a final elevation of 600 metres above sea level at the top of the valley.

Just before reaching the top of the valley there is a pull-out at a pleasant place to stop where there is a nice stone-slab table with pieces of basalt columns for seats. From here there is a nice view over the canyon Fossárgljúfur, as well as into Lambadalur valley, all the way south to Hofsjökull glacier. From this rest stop it is 4.6 kilometres to the lake Ytra-Stafnsvatn, and from there we drive along a grassy area and wetlands in some parts.

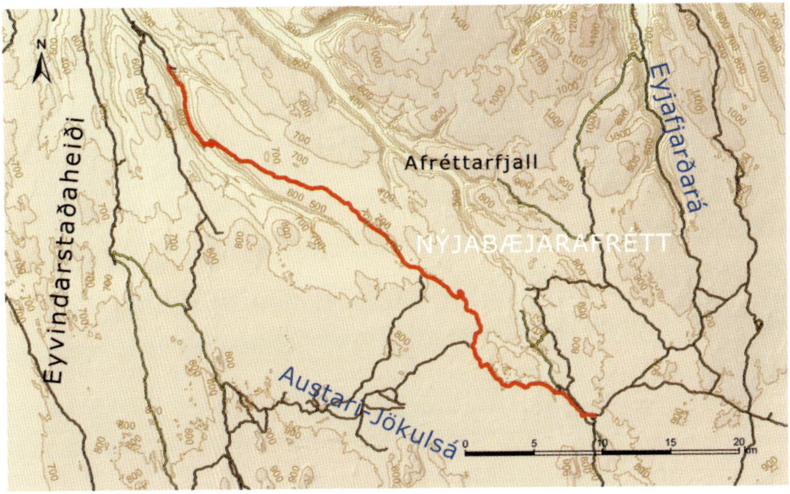

Fording the river Hnjúkskvísl in high water. Photo: Jón G. Snæland.

At the ruins at Langavatn is a very rough track, not taken, which leads down to the valley of Keldudalur.

Just south of this intersection is another track to the right/east at Langavatn that lies along the edge of Þverdalur valley, from where one can also see Keldudalur. However, we do not take either of these options.

At the west slope of Mt. Reyðarfell is an intersection with options not taken, but a sign here points to Ásbjarnarvötn, 17 km away, on the way to Ingólfsskáli hut. The hut nestles under the north slope of Hofsjökull, and is 28 km from the intersection.

South of the intersection, we ford the stream Runukvísl and later arrive at a point even with the west edge of Reyðarvatn lake. Here there is a rough short track to an old hut, Rústaskáli, that is also called Klasar, where there is a horse barn, two half-collapsed turf huts and a shepherds' hut.

South of Rústaskáli, in an area known as Orravatnsrústir, is a track not taken to the west, which is only passable for modified jeeps; it is called Stompleið. It joins a road that lies north of Orravatnsrústir going west to Ingólfsskáli, and as mentioned, this is also only for modified jeeps. Travel beyond Ingólfsskáli is extremely dangerous on these and other tracks, and is ONLY for highly experienced mountain drivers of super jeeps.

We continue straight, going towards Laufafell through a rather barren landscape, with the only vegetation surviving in moist depressions were streams and brooks meander. Soon we arrive at a large glacial river called

An old, collapsed outbuilding in the Orravatns ruins. Photo: Jón G. Snæland.

Austari-Jökulsá, which is one of the main source branches of Héraðsvötn, the large lake that empties into Skagafjörður. Austari-Jökulsá is formed by several branches emerging from Hofsjökull, with another branch to the south emerging from a sub-glacier called Klakksjökull. These branches straddle the prominences of Vesturbugur and Austurbugur, and the river was bridged at the latter site in 1970. This bridge has been transplanted, having previously been over the river also called Jökulsá, but on Sólheimasandur in the south part of Iceland! The Austari-Jökulsá is fed by many branches and is usually a large and intimidating river. It flows down into the Austurdalur valley into Skagafjörður.

Once across the bridge over the Austari-Jökulsá, it is a short distance to a ford across the brook Strangilækur, where the bottom is smooth and reliable, and the stream is shallow during summer. The next stream to consider is Hnjúkskvísl, a glacial stream that can grow rapidly on warm, summer days and become impassable to all but the most substantial superjeeps. Therefore, it is important to examine the ford site carefully before entering the stream.

South of Hnjúkskvísl, the road branches, with one option going northeasterly, to the lawn of the oldest hut at Laugafell. The other option continues a kilometre to the south and up a slope where there is a sign which

This unique resting place comes to light as you come up out of Vesturdalur valley onto the highlands. Photo: Jón G. Snæland.

gives the distance to various destinations: Laugafell 1 km, Eyjafjörður 41 km, Skagafjörður 50 km, and Nýidalur 50 km.

From the sign it is about 200 m down a slope to the east, to a ford at the stream Laugafellskvísl. After the ford it is a short distance to the parking place for the hut run by the touring club Ferðafélag Akureyrar. Here there is a natural pool, perfect to warm yourself at journey's end.

Fording the river Hnjúkskvísl in high water. Photo: Jón G. Snæland.

Skjaldbreiðarvegur

Start–Finish **Kaldadalsvegur, F550 – Kjalvegur, F35.**
Distance/Time **53 km. About three hours' drive.**
Fuel **In Mosfellsbær, at Laugavatn and at Geysir.**
Caution needed due to **Risky fords.**
Passable for **All jeeps.**
Of interest **Glacier views, lava formations and majestic mountains.**

This route is a line road, which is what roads are called that follow high tension lines.

The trip starts with a road to the east (F338), just below the emergency hut on Kaldadalsvegur. Soon a small brook called Sandvatnskvísl is crossed, and then after a short distance we enter the Skjaldbreiðarhraun lava. North of the road rises the prominent glacier Þórisjökull and beyond that, Langjökull spreads out, and the road follows the glacier part of the way.

North of Skjaldbreiður is a road leading to parking place and there is a hiking trail to the edge of the mountain's crater. As the crow flies, the trail is 2.5 km (1.6 miles). Ever since Iceland was settled there have been hiking routes through the highlands, and one such route, Skessubásavegur, is posted south of Skjaldbreiður. A short ways from there and north of the road is Mt. Tjaldafell, at the base of which is a rather tight cluster of huts. Most of these huts belong to various snowmobile groups that take trips onto Langjökull glacier.

A row of jeeps threading the narrow track in along the Brekknafjöll mountains on the way to Lake Hagavatn. Photo: Jón G. Snæland.

The route continues to follow the power line, providing a good road to Mt. Þórólfsfell, a mountain which it is possible to drive around, which is north of Mt. Hlöðufell. It is also possible to drive around Hlöðufell, but the going is tough. There is a hut operated by Ferðafélag Íslands at the southwestern end of Hlöðufell.

Hlöðufell is a majestic mountain, 1,188 metres high, and surrounded by a belt of cliffs. On the summit are communication masts of various types. It is only possible to climb the mountain from the south, ascending from the hut location.

East of Þórólfsfell, we drive up Lambahraun lava, which ran from the volcanic crater Eldborg. North of the road two glacier tongues of Langjökull come into view: Vestari-Hagafellsjökull and Eystri-Hagafellsjökull, both of which plunge into Hagavatn lake. Between them is Mt. Hagafell, which slices deep into Langjökull. The road then skirts the edge of the highlands at Mosaskarð gap, with mountains to either side of the gap. A track is visible to the left, and it follows the mountains of Fagradalsfjall and Brekknafjöll to Hagavatn lake, which is discussed under the section on Hagavatnsvegur (pg. 67). This part is very difficult and only passable by modified jeeps. However, it is easy to walk along the track to the glacial lagoon Hagavatn and see where the glaciers plunge into it.

The 1186 meter high, cliff-bound Mt. Hlöðufell in winter costume. In front of it is a smaller mountain called Þórólfsfell. Photo: Jón G. Snæland.

We continue down the slope from Mosaskarð to Haukadalsheiði heath. Here a monument has been erected to Greipur Sigurðsson, a school principal at Geysir who was also a nature warden and early advocate of soil

Driving up to the Mt. Skjaldbreiður crater is a popular winter trip among Icelanders. Photo: Kjartan Gunnsteinsson.

reclamation on Haukadalsheiði. The track continues to the right, down Haukadalsheiði, a short distance from the monument and down further to Geysir. Just east of the monument there is also a faint, optional track to the left out onto a headland by Sandvatn lake, where efforts at soil conservation are being made. Directly ahead is the river Ásbrandsá, which is forded. If the water is brown, then there is a flood from Hagavatn lagoon and the river is impassable. In this case it is best to turn around and drive down to Geysir.

Skógarmannafjöll

Start – Finish	**Route 1 in Búrfellshraun lava – The hut, Fjallaborg.**
Distance/Time	**12.7 kilometres. Barely an hour's drive.**
Fuel	**Reynihlíð on the shore of Lake Mývatn.**
Caution needed due to	**No particular hazards.**
Passable for	**All jeeps.**
Of interest	**Volcanoes, lava formations and old huts.**

We start our journey on Route 1 in the Búrfellshraun lava, east of Lake Mývatn, 11 kilometres past the intersection to Krafla Power Plant, where we drive south on the flat surface of Búrfellshraun towards the mountains. Búrfellshraun is well vegetated, and the road is both hilly and crooked, so the going is slow. Openings in the lava are seen widely from the road and these can be dangerous. Many side tracks meet our route, some are marked on maps of good scale, others not.

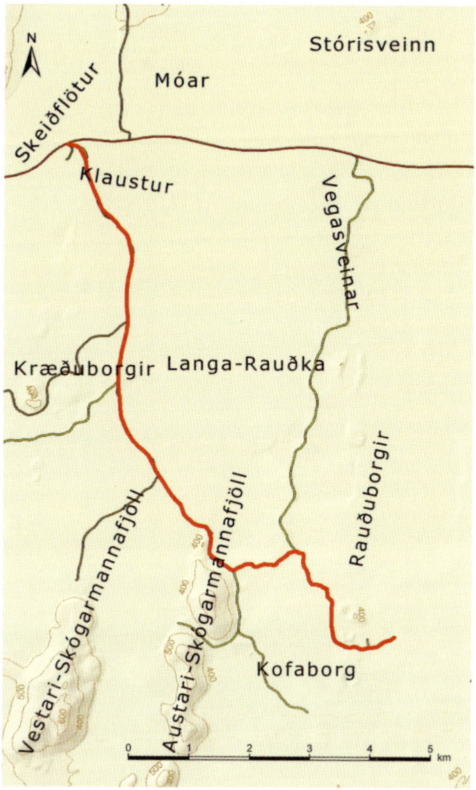

After turning south and travelling 3.4 kilometres there is an optional track to the right. It is a four-kilometre-long track leading southwest to Kræðuborgir, a row of craters several kilometres long from which the Búrfellshraun lava flowed.

A short distance, 1.1 km, south of the junction with the Kræðuborgir track is another track to the right, which leads 16 kilometres west to a hot springs area at the base of Mt. Námafjall, and this track can even be driven in small jeeps. But we continue southward.

We head for the northernmost peak of the Skógarmannafjöll mountains. It is pyramid-shaped and to the right Mt.

The northernmost peak in the Skógarmannafjöll mountains. Photo: Jón G. Snæland.

Olíufjall is visible, and west of it is Mt. Búrfell itself. The northern peak of the Skógarmannafjöll mountains is not marked as a particular point on the map, therefore it can be confusing for travellers using GPS navigation systems. Beneath the mountain, we drive down a steep slope and go south of the mountain. At GPS point: N65 35.729 W16 31.017, turn from

An old shepherds' hut known as Rauðborgarkofi. Photo: Jón G. Snæland.

the mountains, due east into the barrens of Mývatnsöræfi where the road wanders in countless curves through the lava until it finally leads directly south of the volcanic crater Stóra-Rauðka. At the base of Stóra-Rauðka is a shepherds' hut called Fjallaborg, which is half-owned by some individuals in the town of Húsavík, and half by the Round-up Fund of the district of Skútustaðir. The hut is well-equipped and most hospitable. Just southeast of the hut is a track to the old shepherds' hut called Rauðborgarkofi. This hut is stone-built with a corrugated iron roof that is by today's standards rather low. East of Rauðborgarkofi is the lava flow Nýjahraun, which Ari Trausti Guðmundsson describes in his book *Íslenskar eldstöðvar* as having flowed during the eruption of Askja in 1874–1875, when the crater Sveinagjá also erupted and was known locally as *Sveinagjáareldar* (the Fires of Svein Crater).

Snæfellsleið

Start – Finish	**Fljótsdalur valley – West of Eyjabakkajökull glacier at Maríutungur.**
Distance /Time	**74.5 km. Two to three hours.**
Fuel	**At Hallormsstaður.**
Caution needed due to	**Snow drifts can persist into July as far in as Maríutungur.**
Passable for	**All vehicles during summers.**
Of interest	**Majestic mountains, natural thermal pools, glaciers, and reindeer.**

This trip starts with a turn west off Route 931 deep on the west bank of Fljótsdalur valley at a tumultuous chasm on the river Bessastaðaá. Here, on Route 910, which is called Kárahnjúkavegur, we zig-zag up a very steep slope to the Fljótsdalsheiði heath at about 600 metres elevation. The road is raised and paved all the way south to Mt. Sauðafell. Once on the heath, a new section of rather unexciting road begins to lead us towards Mt. Snæfell.

As we approach the first mountain, called Laugarfell, and before the road bends to the west, there is an optional turn-off to the left. This leads to the river Laugará and a shepherds' hut near a natural thermal pool. From the Laugará we follow the base of Mt. Laugarfell, where there is a paved

At 1833 meters above sea level, the ancient central volcano Mt. Snæfell is the highest freestanding mountain outside the glaciers. Photo: Jón G. Snæland.

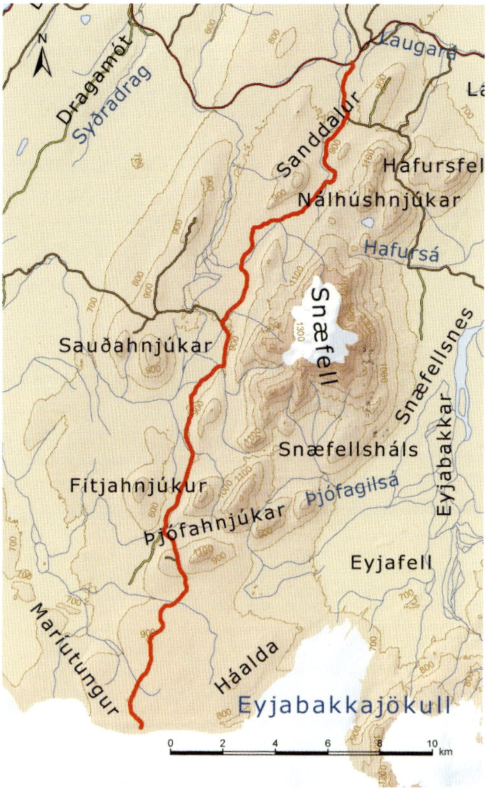

section west of the mountain, which leads to the Landsvirkjun dam area at Eyjabakki. Shortly after this is a track that leads to the east of Mt. Sauðafell and then south around it to the main route inland toward the Snæfellsskáli hut in the valley Sanddalur. This is a difficult stretch, but the main part is extremely beautiful, especially if driven further in along the prominence Nálhúsahnjúkar.

The main route in along Snæfell lies just to the west and is fairly easy going, but we do have to cross a few unbridged brooks in the process.

West of Mt. Snæfell is a hut named Snæfellsskáli, managed by the touring club Ferðafélag Fljótsdalshéraðs. This is where the main trail is, leading to the top of Mt. Snæfell, which is the highest individual mountain outside of glaciers in Iceland, and has an elevation of 1833 metres above sea level. Snæfell is Icelandic for 'Snow Mountain' and ironically it is an ancient volcano.

Standard vehicles can go no further than Snæfellsskáli, but jeeps can continue west from the hut a short distance, to where you can hike up Sauðahnjúkar, where there is a spectacular view across the barrens of Vesturöræfi, which is prime habitat for Iceland's reindeer population.

Inland past the hut is a track southward, leading to Maríutungur nestled between the foot of the glacier Brúarjökull to the west and another smaller glacier, Eyjabakkajökull, to the east. Soon after leaving the hut, we need to cross a river, which is one of the source branches of the river Sauðá. The track is a bit vague where the river should be crossed on gravel bars, but it becomes more obvious across the river. At Mt. Bjálfafell the road is

Nálhúsahnúkar (Needle-storage spikes) to the north of Mt. Snæfell. Photo: Þóra Sigurbjörnsdóttir.

at an elevation of 1000 metres and the view here is gorgeous. Many small brooks are crossed on the way towards Eyjabakkajökull, but mostly on a good track, practically all the way to the glacier where there are sizeable ice caves, and our journey ends.

A thermal pool at Laugarfell. Photo: Jón G. Snæland.

Sprengisandsleið

Start – Finish	**Hrauneyjar – Aldeyjarfoss waterfall.**
Distance/Time	**205 km. A day trip.**
Fuel	**At Hálendismiðstöð station at Hrauneyjar and at Fosshóll near Goðafoss waterfall.**
Caution needed due to	**Fords at Nýjadalsá river and at the branch Hagakvíslar.**
Passable for	**All 4x4 jeeps, but not SUV's.**
Of interest	**Great views, including of glaciers, waterfalls.**

We start our trip at the site known as Hald, as few kilometres south of the Hálendismiðstöð Highland Centre at Hrauneyjar. Here there is also a short track to the Tungnaá river, where a small ferry, also named Hald, was operated. At the ferry site we can see stone fences used to hold sheep awaiting transfer across the Tungnaá in small boats. Farmers still herd their sheep to highland pastures to roam free until fall round-ups. There is a monument at the Hald ferry point honouring Sigurjón Rist, an hydrologist and early advocate of highland travel in Iceland.

This ferry was important in the first automobile trip across Sprengisandur in 1933. Five Icelanders were involved and they used a Ford Model 1925

The mountain Syðri-Háganga by the Hágöngulón reservoir just east of the Sprengisandsleið road. Photo: Þóra Sigurbjörnsdóttir.

that was ferried by cable-assisted boat across the Tungnaá. North of Hald is a platform cable-ferry that was used to transfer sheep and smaller vehicles across the river, before the newer bridges were built.

A few kilometres to the north is the Hálendismiðstöð Highland Centre at Hrauneyjar, which is the last place to buy fuel before entering the highlands. The next place to buy fuel is not until Bárðardalur valley in the north.

From Hrauneyjar we continue north to Mt. Vatnsfell, where the gravel road starts and the highland sensation takes over.

Sprengisandsleið (F26) is Iceland's best-known highland route, and has been for centuries, even though travel across it ceased at intervals over the ages. For a while it was due to people's fear of outlaws, when they did not travel Sprengisandur unless heavily armed or in large numbers. In 1772 Einar Brynjólfsson and four others headed north over Sprengisandur in the first trip made there in 30 years. By pure chance they happened upon the outlaw couple Eyvindur Jónsson (Fjalla-Eyvindur) and Halla, his partner, and the couple was captured. Another reason people avoided travel across Sprengisandur was stories of people getting lost or stranded by weather in the highlands.

Shortly past Vatnsfell we see a large attractive body of water which is called Þórisvatn and is in part a reservoir which has a good bit of trout in it. Farther on we arrive at a site known as Stóraver, where we turn right off the main road, to a hut called Versalir at Stóraversskurður. Several years

The Lindarkot lodge at Þóristungur, which is an oasis of vegetation, above which rise rather grand but bleak and bare high swells. Photo: Jón G. Snæland.

ago, tourist services were provided here during the summer. At Versalir, we cross Stóraversskurður and Route F26 starts on the other side of the bridge. The section of road from Versalir into the valley of Nýidalur has been called *Ölduleið*, due to the many large swells along the way. The first of these is called Þveralda, where the road winds through, and from which there is a good view. North of Þveralda is the river Svartá, which is crossed at a ford where much caution is needed because the bottom is irregular.

The road lies west of another swell, Hnöttóttaalda, and along Hnausaver in towards Skrokkalda and Hnausalda. At the base of Hnausalda, the stream called Hnausaverskvísl runs out of a canyon where there are beautiful beds of the wildflower Arctic River Beauty, perhaps the most impressive of such beds in the country. Just north of Skrokkalda is a decent jeep track leading to the reservoir Hágöngulón. From Skrokkalda we drive over a few harmless streams north towards Nýidalur valley.

Ferðafélag Íslands has a guest hut in Nýidalur and it is possible to stay either there or in the adjacent campground. Nýidalur is at the base of the glacier Tungnafellsjökull, and has a glacial river in it called Nýjadalsá. A popular hike is from Nýidalur over to Vonarskarð pass where there's a beautiful view over the pass and across to the mountain Bárðarbunga and

Vatnajökull. Many also hike down to Stapadalur valley to see the unique hot springs area there.

From the hut, we drive over the Nýjadalsá, which as a glacial river can have a widely varying amount of water in it. Usually there is not enough water in the river to make it impassable, except in wet weather or hot summer days. Therefore, travellers should get advice from the hut supervisor about the river, if they are insecure about crossing rivers at fords. A short distance north of the Nýjadalsá, and similar-sized, are the streams Hagakvíslar in Tómasarhagi, but these have more current. Tómasarhagi is named for Tómas Sæmundsson who got lost on his way south across Sprengisandur in 1835 and found this small grassy area. Right at the ford of the Hagakvíslar is the turn-off to Gæsavatnaleið, a route also discussed in a section of this book (pg. 62).

As we continue, a high swell known as Fjórðungsalda rises in the north and at the foot of which is the lake Fjórðungsvatn that nearly dries up in summers. We drive past a route that leads to the west, discussed in another section of this book (pg. 58).

An expanse of Arctic Riverbeauty to the south of the Skrokkalda swell right beside the Sprengisandsleið road. Thought to be the largest such expanse in Iceland. Photo: Jón G. Snæland.

Farther north on our way, at a point even with the location of Íshólsvatn lake are the cliffs of Hrafnabjörg, where it is interesting to stop and walk to the edge of the cliffs looking down over the vast lavas of Ódáðahraun, and the waterfall Hrafnabjargafoss, as well as down to the river Skjálfandafljót, all of which are below the cliffs. As we descend out of the highlands we reach the parking place at Aldeyjarfoss waterfalls, one of Iceland's most beautiful waterfalls, and the endpoint of our journey. A few hundred metres below Aldeyjarfoss is a smaller, lesser-known waterfall, Ingvararfoss.

Veiðivatnaleið

Start – Finish	**Vatnsfellsvirkjun Power Plant – Tjaldvatn lake (at the hut cluster there).**
Distance/Time	**21 km. About 40 minutes' drive.**
Fuel	**At the Hálendismiðstöð station at Hrauneyjar.**
Caution needed due to	**Fords.**
Passable for	**All jeeps during summers.**
Of interest	**Beautiful lakes, volcanic craters and fissures, panoramic views and desert.**

We drive north of the Hálendismiðstöð Highland Centre at Hrauneyjar and then right/east at the intersection where F26 meets F228, below a large hydro-electric dam named Vatnsfellsvirkjun. It may be interesting to first drive another kilometre on F26 to see the dam, where there is sometimes a waterfall when reservoir overflow spills over the dam. Veiðivatnaleið traverses the reservoir area and goes over a rise south of Mt. Vatnsfell and on a raised road down to the lake Fellsendavatn, where the actual mountain track starts. The track is sandy and we drive more or less on coal-black sand in to the Veiðivötn lakes. Therefore, there is a risk of sandstorms any time of the year. There are a few detours along the way, but few are posted so it can be difficult to navigate, unless there is a GPS system in the vehicle

At the Veiðivötn lakes there is a sizeable cluster of fishing lodges and a travellers' centre. Photo: Soffía Snæland.

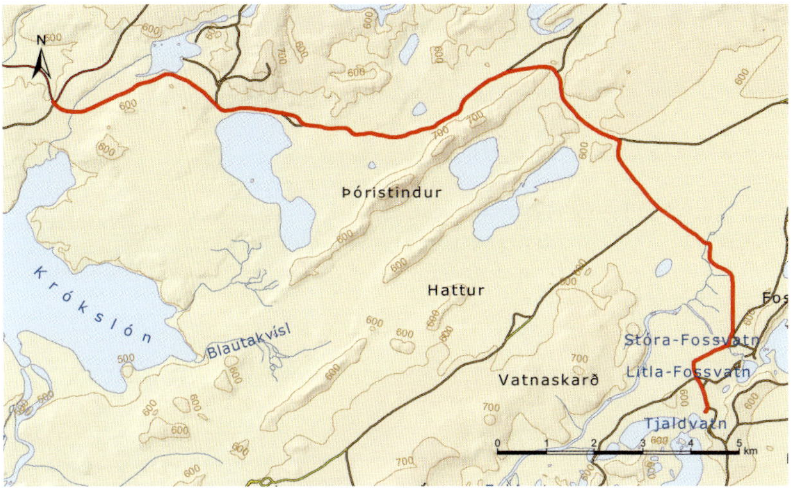

and you have a map published in Iceland. The reason is that sand quickly covers the tracks from the previous driver.

The first hindrance along the way is a ford at the stream Vatnakvísl, which is however passable by all 4-wheel drive vehicles during the summer. The ford bottom is a bit irregular, but it is obvious how to cross. It is a bit

A lone fisherman trying his luck in one of the many lakes at Veiðivötn. Photo: Steingrímur Steinþórsson.

Arnarpollur in the Veiðivötn area. Photo: Steingrímur Steinþórsson.

of a steep climb out of the stream to the east. Next we come to the stream Fossvatnskvísl, where there is an intersection. The option not taken to the left leads northeast all the way to Tungnaárjökull glacier and a hut there called Jökulheimar.

Another option leads to the lake Stóra-Fossavatn. Just above the ford is a pretty waterfall in the stream, where it exits Litla-Fossvatn lake. The ford on Fossvatnakvísl can be too deep for small jeeps, so a small turn is taken in the lower part of the ford, where it is shallower. Larger vehicles can ford going straight across, though. From the ford it is barely two kilometres to the cluster of huts at Tjaldvatn lake.

The main hut here is called by locals *Sæluhúsið í Veiðivötnum*, and below it is an old turf-built hut that has been newly restored. Those needing accommodation or information about the area should visit Rúnar or Bryndís at the service hut Varðberg.

Veiðivötn is a cluster of about 50 lakes, many of which are in fissures or craters formed during an eruption of 1477 along a 67-kilometre-long fissure. There is considerable trout fishing in the lakes and in previous times, farmers made special trips to the Veiðivötn (meaning, 'Fishing Lakes') to catch Arctic charr (*Salvelinus alpinus*) and brown trout (*Salmo trutta*).

From Tjaldvatn lake, tracks lead in various directions. One is called

Tröllið (The Troll). Tungnaá River in the background. Photo: Steingrímur Steinþórsson.

Vatnahringur, which starts past the service hut Varðberg, and up the slope above the lake. It then circles the southern collection of Veiðivötn lakes, down to one known as Snjóölduvatn, and winds between more lakes and returns to close the circle on Veiðivatnaleið leading back to Varðberg. This is a beautiful route, so you should give yourself at least an hour to do it. The only noteworthy hindrance is a ford on the stream Kvíslavatnskvísl, which is somewhat deep, but has a reliable bottom.

It is also possible to drive a larger lake circle, which starts south of the service hut Varðberg. Another option is to drive the trail to Tröllið, a large stone spire south of the mountain ridge Snjóöldufjallgarður, above the Tungnaá river.

Þakgil

Start–Finish	**Route 1 at Höfðabrekka – Þakgil at Höfðabrekkuafréttur high pasture.**
Distance/Time	**15.5 km. One hour's drive.**
Fuel	**In the village of Vík í Mýrdal.**
Caution needed due to	**Creeks and streams.**
Passable for	**All jeeps, including small ones.**
Of interest	**Caves, cliffs, glaciers, glacial rivers, and waterfalls.**

The route into Þakgil, or *Höfðabrekkurafréttur*, as the whole area is called in Icelandic, starts from Route 1, at Hótel Höfðabrekka, east of the village of Vík í Mýrdal.

There we enter the valley of Kerlingardalur, and climb a steep slope near the farmsite Reynisbrekka, and continue on to the high pasture. This part of the route and into the area of Múlakvísl (Sandvatn) is part of the old Ring Road around Iceland.

At Múlakvísl stream there are traces of the old bridge that was destroyed by a burst flood following the sub-glacial volcanic activity of Mt. Katla in 1955. But before we arrive at Múlakvísl, we drive along a heath with fascinating sandstone formations in the cliffs, as well as several caves.

Þakgil gorge provides magnificent natural scenery, waterfalls, cliffs, precipices and many caves. Photo: Soffía Snæland.

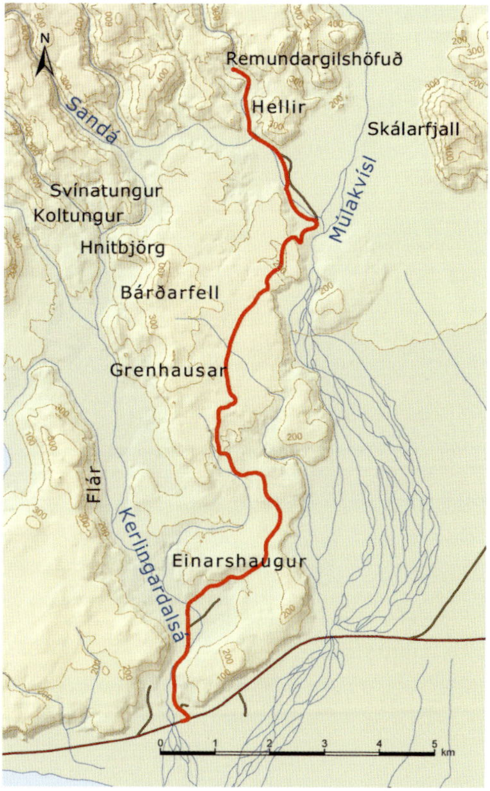

At a point even with the site known as Léreftshöfuð, we drive down a steep slope and across the stream Múlakvísl on a culvert that is several years old. The name of streams here can be confusing. Sandvatn is a large glacial river, and not a lake as its suffix might imply. The smaller stream is Múlakvísl, but upstream it is called Afréttará. Where Sandvatn and Múlakvísl join, the resulting river is called Múlakvísl taking the name of the smaller tributary which is unusual.

From the culvert we drive over a high gravel rise and soon meet a track branching to the right and leading to a pretty chasm called Remundargil. Also, off of this track, is another track leading all the way to Sandvatn. But we continue straight on.

We pass below a bluff called Hvolshöfuð, and then between Hvolshöfuð and Þakgilshöfuð, where we cross a few harmless brooks. Also visible is a track following the Afréttará to the guest hut called Ausuból.

Between the bluffs is the site of Miðfell, where there is a cave that was used until 1918 by shepherds herding sheep from the high pasture. Many have scratched their names or a date on the soft sandstone of the cave walls, with the earliest date being 1804. Travelling inland from the cave, a view into Þakgil opens, with border bluffs 100-200 metres high and cliffs in all directions from the narrow canyon walls. Being narrow, the canyon offers good shelter and some tourist services are available. A campsite with sanitary facilities was opened in the year 2002 and since then some small cottages have also been built and made available for rent.

But what raises the most visitor interest at Þakgil is the cave at the camp-

Above Þakgil looms the glacier Mýrdalsjökull with its glacial tongues creeping down through gorges and gullies. Photo: Kjartan Gunnsteinsson.

site. The couple Bjarni Jón Finnsson and Helga Ólafsdóttir of Vík have made many modifications to the cave, with table and seating for a couple of dozen people. Candles have been installed on the walls and there is a

The Þakgil camping grounds with beyond it a cave furnished with a table and benches which has proved very popular with Icelandic travellers. Photo: Soffía Snæland.

cooking grill as well. Two home-made ovens made from metal buoys allow wood scraps to be burned for heat.

There are many hiking trails in the area and there is a route from the outer reaches of Þakgil up to the glacier Mýrdalsjökull. We then return over Miðfell past that cave, where there is a very steep and narrow track that zig-zags up towards the glacier. The track is very narrow, and very close to the edge at times, so those with a fear of heights are not advised to try it. This a six-kilometre track leading to Sker and Jökulhöfði. Another track leads a short way east to Rjúpnafell. There is a unique view from there over the glacier and across to Hafursey.

Þingmannaheiði

Start – Finish	**Vatnsfjörður on Route 60 – Skálmarnes on Route 60.**
Distance/Time	**24 kilometres. About two hours' drive.**
Fuel	**At Flókalundur.**
Caution needed due to	**Large boulders, runoff and small rivers.**
Passable for	**Jeeps, preferably with a tyre size of at least 33 inches.**
Of interest	**Ancient cairns, rugged and majestic wilderness.**

We drive into this route just east of Flókalundur, south of a bridge over the Vatnsdalsá river. There we drive north along the river Þingmannaá which courses in pretty canyons through dense shrub-land.

The smallest jeeps only manage a few hundred metres of this since the track becomes quite rocky and rock-slabbed in the very first slope. Along the way are many picturesque cairns that were built in 1896 under direction of the local politician Guðmundur Jónsson. A letter from him to a higher district official, dated July 15, 1896, states:

> Upon the recommendation of the District Committee Member, Father Þorvaldur Jakobsson of Brjánslækur, I have taken upon myself the task

Driving over the very rocky terrain south of Djúpavatnshorn can be quite a struggle. Photo: Jón G. Snæland.

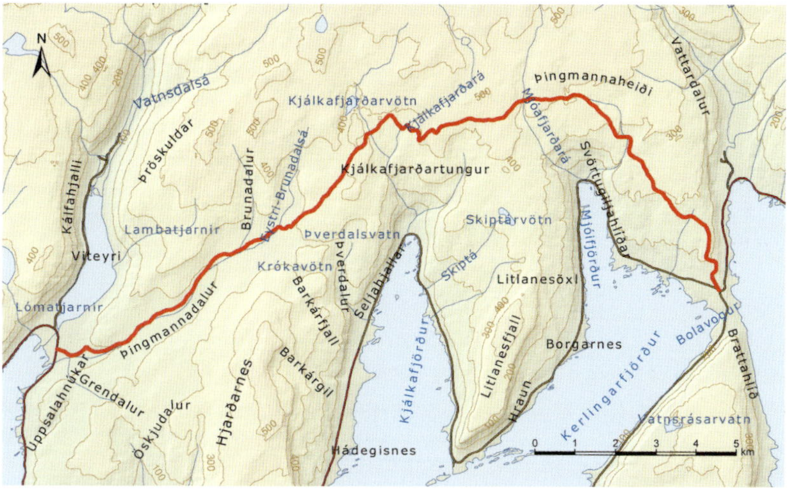

of road-grading and cairn-building along the public road across Þingmannaheiði heath, which lies between Barðaströnd and the district of Múlahreppur, for 500 krónur.

This work was done in the months of June and July this year, all the way from Smiðjukleifar in Vatnsfjörður fjord, to the so-called Djúpavatn south of the heath, and the work involved: removing rocks, gathering and shovelling smaller rocks from the roadway, and raising the surface

The road leads up out of Vatnsfjörður along the river Þingmannaá and follows an old cairn-marked route over the heath. Photo: Jón G. Snæland.

The Hellulaug hot spring by Flókalundur. Photo: Jón G. Snæland.

> of the road with gravel in certain places where it was felt most needed, adding fill to the road, moving it in many places – both to remove unnecessary turns and because of water etc. – 125 cairns were built, since there were previously none in the area from Smiðjukleifar south to the so-called Grettistak.

The accompanying receipt details that 150 days' work was done and that the salary for road workers was three krónur a day.

There are also posted hiking routes along Þingmannaheiði, with posting between the cairns wherever it was deemed necessary.

After coming up the first slope we drive along Þingmannadalur. Farther up in the valley is an intersection with an optional track to the south down to the valley of Fossárdalur. This route was driven until 1974 such that the west part of Þingmannaheiði heath was driven until then. Just east of the Fossárdalur road is another track to the north, which is a line road that leads down to the hydro-power plant Mjólkárvirkjun. Those taking that route can divert from it midway and go onto the Dynjandisheiði heath. At the highest point of the track, called Kjálkafjarðartunga on maps of the Icelandic Geodetic Survey, the road is at an elevation of 414 metres above

One of the old cairns in the Þingmannadalur valley on Þingmannaheiði heath. Photo: Jón G. Snæland.

sea level. Here there are two huts, one is a supply shack for the West Fjords Energy Utility, Orkuveita Vestfjarða, and the other is an emergency shelter with the main necessities, including a bed, heating equipment and food in vacuum packages. Those staying in this hut can expect a visit from the ghost of Rassbeltingur, who is the same spirit known as Hringdalsdraugur and Brjánslækjardraugur. According to sources at the Ghost Centre in Stokkseyri, this is the ghost of a man who froze to death on Þingmannaheiði.

Just east of the huts we drive across one of the source branches of the river Kjálkafjarðará. We take a ford across this river and all others along our way, since all culverts and bridges along the route have collapsed. A bit to the east the route curves southward down to a point between Mjóifjörður to the west and Vattarfjörður to the east. At this point there are no more cairns. The route ends at the site Eiði on the Skálmarnes peninsula.

Þingmannaheiði was first driven by auto in 1949 by the pioneer mountain tour leader Guðmundur Jónasson. In 1950-1951 Þingmannaheiði was graded and traffic began during summers after that. In the years to follow, roadwork continued in the district of Austur-Barðastrandarsýsla, with residents of this area being linked to the national road system in 1951.

Use of Þingmannaheiði was then discontinued in 1969 when the new highway along the shoreline was made. The exception was the previously mentioned part that lies west and down to Fossháls and further to Foss, which was used until 1974. This section is quite rough and barely passable for jeeps. With the coming of the shoreline road, the distance between beginning and end points lengthened by 11.9 kilometres. But, despite the fact that Þingmannaheiði zig-zags considerably, it would be possible to shorten the distance between points by at least 15 kilometres by renewing the road along Þingmannaheiði.

Þórsmerkurleið

Start – Finish	**The old bridge over Markarfljót River – Pedestrian bridge over the river branch Tungnakvísl.**
Distance/Time	**About 24 km. Barely two hours.**
Fuel	**At the town of Hvolsvöllur.**
Caution needed due to	**Glacial rivers.**
Passable for	**All jeeps during summers.**
Of interest	**Glaciers, glacial rivers, caves, canyons, waterfalls and mountains.**

An eruption began on Fimmvörðuháls on March 21, 2010 and continued until April 13, 2010. Lava from this crater ran down the canyons of Hrunagil and Hvannárgil causing a lava fall of 200 metres height – considered the highest lava fall to occur in human history. On April 14, 2010 another eruption started, this one in the top crater under Eyjafjallajökull, which last erupted intermittently during 1821–1823.

Soon after the current eruption started, a glacial flood burst descended along Gígjökull and onto the gravel flats of the river Markarfljót. As the heat at the summit melted glacial ice, there were explosive eruptions generating incredible volumes of debris – as much as 750 tons per second – which spread throughout Europe in the jet-stream. This caused countless

By the eruption site on Fimmvörðuháls ridge, winter 2010. Photo: Bæring Björgvinsson.

flight cancellations for days to follow, due to the risk to jet engines caused by ultra-fine ash.

To reach Þórsmörk, we turn off Route 1 at the waterfall Seljalandsfoss and drive north on Route 249 in to the old bridge over the river Markarfljót. Here there is a single building and a parking area. This is where those with smaller vehicles leave them and take a bus into Þórsmörk. Some may be fascinated by the old bridge, which appears to be on the verge of col-

Driving across the Krossá river by the Merkurrani. Photo: Jón G. Snæland.

Working on soil reclamation on the Merkurrani in Þórsmörk. Photo: Þóra Sigurbjörnsdóttir.

lapsing. But nonetheless it withstood the post-eruption floods of 2010 and was used to transport milk across the river when Route 1 was closed due to the eruption floods. In order to stay within weight limits, the milk was shuttled over by trucks having smaller tanks, going between two larger tank trucks.

It is also fun to study the churning waters under the bridge and imagine what it must have been like in the early days when men like Vatna-Brandur operated trips east to the village Vík í Mýrdal, crossing the river in a 2-wheel drive, Model 1927 Ford.

But the true start of the Þórsmerkuleið is at this parking place and from here our route takes us on a rough track through the gravel flats of the river Markarfljót; rough because it is made of the very gravel surrounding us.

On the way in towards Gígjökull glacier, we cross many brooks none of which are of any concern during summer.

The first large ford used to be across the river Jökulsá from Gígjökull (Glacial River from Gígjökull). It is has been variously crossed either at the lagoon beneath the glacier itself, or down on the gravels. But the lagoon is gone due to the eruption in 2010 and so is the river.

The next river to cross is Steinholtsá, which is a glacial river with a tendency to change its channel, and is variously in one or two branches. It can grow quickly and have quite a stiff current and be hazardous, so it is

The explosive ash and tephra eruption in the top crater of Eyjafjallajökull, spring 2010. Photo: Bjarni Ingibergsson.

important to examine the ford carefully each time it is crossed. It can be a good idea to have another vehicle nearby when fording.

We reach an intersection beyond Steinholtsá, with an optional turn to the left going down the gravel flats toward the valley of Húsadalur, where Hostelling International Iceland operates accommodations and various other visitor services. On the way, the river Krossá must be forded at the most difficult ford in Þórsmörk. The river is constantly changing its channel and therefore a proper place to ford must be found each time you cross. The ford has often been up against some cliffs there, but it sometimes changes when the river decides to wander onto the gravels. Under no circumstances should the Krossá be forded without another vehicle nearby.

But continuing straight ahead to the river Hvanná and across it. Hvanná has a stiff current, running along rocky gravel bars, and often follows along the road for a goodly distance, after the ford itself was graded. However, it can be said that the Hvanná is the friendliest river in Þórsmörk because at least it is clear, while the others are glacial. Nevertheless, the Hvanná can grow quickly and carry with it large boulders into the ford, which then becomes hazardous to vehicles.

Having crossed the Hvanná we arrive at another intersection. The track

A view from Merkurrani to the Tindfjöll mountains. Photo: Steingrímur Steinþórsson.

to the left goes over to the valley Langidalur and the hut operated by Ferðafélag Íslands. On this route you must also cross the Krossá on one of the most heavily travelled fords in Iceland. Jeep drivers tend to come here almost with the express intention of meeting the challenge of crossing the Krossá, which as mentioned, constantly changes its channel.

On the other hand, if we drive through the above-mentioned intersection, and in to the area known as Goðaland, we soon arrive at the hut, called Básar, run by the touring club Útivist. From Básar it is possible to continue 3.5 kilometres to a pedestrian bridge over the Krossá river, and our drive ends here.

Near the bridge is a chunk of rock, weighing several hundred tons that fell on the track a few years ago, and it is possible to see where the old track lies under the rock. From this point it is possible to hike to a couple of sub-glaciers: either Krossárjökull or Tungnakvíslarjökull. It should be mentioned that there are campgrounds and small shops at both Húsadalur and Langidalur, and it is also possible to book accommodation in the huts at Húsadalur, Básar and in Langidalur.

Öldufellsleið

Start – Finish	**Route 209 at Hrífunes – F210 at Mælifellssandur.**
Distance/Time	**37.7 kilomtres. 1.5 hours**
Fuel	**At the town of Kirkjubæjarklaustur.**
Caution needed due to	**Small creeks and streams.**
Passable for	**All jeeps and vehicles with high clearance.**
Of interest	**Glacial rivers, waterfalls, glaciers, deserts and mountains.**

Many consider Öldufellsleið one of Iceland's most beautiful drives. We turn off Route 1 at Laufskálavörður on Mýrdalssandur, east of the town of Vík. There, many cairns have been raised just off the north side of the highway, where there is a parking area. Then drive on Route 209 and turn onto to the highlands road (F232) just west of the glacial river Hólmsá. Soon a bridge is reached over another glacial river called the Leirá, which drops in falls and cascades under the bridge and is usually coloured with sediment and somewhat intimidating to see. It is enjoyable to view the river a while from the bridge, and likewise as the road follows it a short distance. On the other side is the other large glacial river Hólmsá, which runs from

A lovely waterfall in the river Bláfjallakvísl where it runs down into the valley Öldudalur. Photo: Jón G. Snæland.

the glacier Torfajökull. Recently a company called Bátafólkið (Boat People) has offered exciting white-water rafting down the rivers.

Once we have driven seven kilometres along the track, we arrive at an intersection on a flat sand plain. Though not part of this route, it is here possible to drive inland along a seldom-travelled track to an old hut called Atley, after a 335 metre high crest of the same name. The hut has been in disuse for a long time and has dry-stone walls and a timber attic. The roof rafters are wooden, with corrugated iron roof covered with sod. Inside the hut has two storeys, with the upper level intended for shepherds and the lower level for their horses or sheep.

However we continue straight and cross the river branch Jökulkvísl on a substantial bridge, and then cross the river Ytri-Bláfellsá, also via a bridge. Then it is a short distance to the Hólmsá river, which we follow upstream until we arrive at the shepherds' hut at the waterfall Hólmsárfoss on the river. The hut is called Framgiljaskáli and is similar to the Atley hut.

Framgiljaskáli is still used for overnight stays and is well-maintained. Just uphill from the hut is a shipping container that has been converted for use as a WC. Hólmsárfoss is a beautiful, large waterfall near the hut. Barely two kilometres north of the hut, a difficult track branches off to the right, leading upstream along the Hólmsá River, which skirts east of the prominence known as Einhyrningur.

From this last intersection, the track leads uphill to the mountain, Loðnugiljahaus, from which it is a short distance over to Öldufellsjökull, one of many glacial tongues that descend from the larger glacier Mýrdalsjökull.

The valley Öldudalur below Mt. Öldufell. In the centre of the picture the top of Mt. Mælifell is visible with the Bláfjöll mountains to the right. Photo: Jón G. Snæland.

We continue driving along an easily traversed cinder field southwest of Loðnugiljahaus and head for Mt. Öldufell. Here the track is on the margin of the valley Öldudalur, a cliff-bordered valley between the road and the glacier. The route then arrives at a ford on the branch Bláfjallakvísl, which is a clear-water stream and only 20 cm deep (8 inches) during summers. The ford is graded when the Road Administration opens the road for the summer. Other streams in the area are only the creek Öldufellslækur, somewhat to the north, and a few harmless brooks.

Just north of the ford over Bláfjallakvísl is a track to the left, leading down in the direction of the glacier. If the track is followed a short distance, we arrive below the ford on Bláfjallakvísl, where there is a pretty waterfall in the river.

Otherwise, we continue along the cinder flats to the north of Bláfjall with Kerlingarhnúkur on the right-hand side. South of the lava crest Háalda, the Öldufellsleið meets the southernmost part of Fjallabaksleið syðri, coming from the right (west). This junction is very subtle and Fjallabaksleið syðri here is little more than a set of tyre tracks, and alertness is necessary to avoid missing it.

Our route continues north of Háalda, which is actually the southernmost part of the Eldgjá volcanic fissure that erupted about 930 AD. The cinder here is likely from Eldgjá, but the volcano Katla could also have sprayed the area with debris. Opposite the middle of Háalda, we enter a sandy lava area where we meet a track from the right (east) leading all the way over to Álftakvísl stream. From here it is only a few kilometres to the uppermost reaches of the Hólmsá river where our route ends.

Öskjuleið

Start – Finish	**Route 1 – Askja**
Distance/Time	**A bit more than 100 km. Two to three hours' drive.**
Fuel	**In Möðrudalur and at Lake Mývatn.**
Caution needed due to	**Two fords along the way: at the rivers of Grafarlandaá and Lindá.**
Passable for	**All 4x4 jeeps, but not SUV's.**
Of interest	**Volcanic crater, glacial rivers, desert, mountains and natural thermal pools.**

We turn south off Route 1 a few kilometres west of the river Jökulsá á Fjöllum (F88). There we pass a volcanic fissure called Hrossaborg, and then a sandy desert to Miðfell. In the southern part of Miðfell is an old hut, called Tumbi, made of stone walls. It is fairly typical of shepherds' huts in Iceland. Sheep and horses are usually let loose to roam free in this area all summer, until they are rounded up in the fall. Round-ups were often difficult, since autumn weather in this region is unpredictable. There are many stories telling of shepherds' incredible ordeals.

South of Miðfell, we come to a small clear-water river, the Grafarlandaá. There is a small but pretty waterfall in it, just upstream of the ford site. The ford is neither deep nor in stiff current, and there is always about the same amount of water in the river. We drive directly across the stream. There is another river further south, called the Lindá, and at the ford site is an information sign about how to drive across the river. The Lindá is

A waterfall in the river Grafarlandaá, just above the ford. Photo: Jón G. Snæland.

larger than Grafarlandaá, but the current is slow. The Lindá is also a clear stream, but sometimes in warm weather, volume in the Jökulsá á Fjöllum can be so much, due to melting of the glacial source, that it floods over into the Lindá, which is then brownish in colour. If this is the case upon arrival, travellers should wait until an experienced Icelandic mountain driver comes along. This can sometimes be a lengthy wait, but in high-water conditions, the Lindá can actually be impassable for smaller jeeps.

South of the Lindá are the springs at Herðubreiðarlindir, a green oasis in the large desert surrounding it. Here the touring club Ferðafélag Íslands has hut accommodation, a campground and information service. Just west of the main hut, is the old hide-out of one of Iceland's most famous outlaws, Eyvindur Jónsson, usually called Fjalla-Eyvindur. Fjalla-Eyvindur spent about 20 years dwelling as an outlaw in the highlands of Iceland. West of the springs, the majestic mountain Herðubreið rises to 1682 metres above sea level. Some call Herðubreið the Queen of Icelandic Mountains. On top of the mountain is a small volcanic crater. It is a popular climb to the top, which takes about four hours. South of the Herðubreiðarlindir springs, the

Víti by Lake Öskjuvatn is a popular bathing place. Photo: Jón G. Snæland.

road lies essentially along the banks of the large river Jökulsá á Fjöllum, which has its source in the glacier Dyngjujökull, a large northern branch of Vatnajökull, Europe's largest glacier. At several points it is possible to stop and view the torrential river.

The route now traverses a cinder-strewn landscape, and under the mountain Herðubreiðartögl, a track is visible to the right, where climbers start their assent. From Herðubreiðartögl, the road bends to the west along an area known as Vikursandur, where fissures and small craters, like the red and black-coloured Höttur, are widely visible.

Ahead are the Dyngjufjöll mountains, a mountain range that surrounds Öskjuvatn lake. Öskjuvatn is part of a geologic formation that was created by a volcanic eruption in 1875. Before the eruption, the lake was only 200 metres in diameter, but now it is 11 km^2 in area and 217 metres deep. At Dyngjufjöll we arrive at Drekagil, where Ferðafélag Íslands has huts. There are now good sanitary facilities and accommodation here.

At the hut, called Dreki, is an intersection. Continuing past the hut takes us southwest into Gæsavatnaleið, Route F910. But above the hut lies the track up to Öskjuvatn lake, ending at a parking area at Biskup (Bishop).

The crater Höttur on the Vikursandur sands just east of Drekagil gorge. Photo: Jón G. Snæland.

From here it is a 30-40 minute walk to the crater Víti. Víti is about 150 metres in diameter and it is about 60 metres from the crater's edge down to the water level. Because the water is very warm, Víti is a popular bathing place. It is best to stay close to the shore, though.

Common endings

-afréttur high pasture
-alda swell
-á river
-brekka, -ur slope, -s
-dalur valley
-ey island
-fell mountain
-fjall mountain
-fjörður fjord
-fljót river
-foss waterfall
-gil gorge
-gjá gorge
-gljúfur canyon
-háls ridge
-heiði heath
-hellir cave
-hnjúkur peak
-hraun lava
-hreppur rural district
-hver hot spring
-höfði cape
-jökull glacier
-kofi hut
-kvísl branch
-leið road, track
-laug pool
-lækur creek
-lón lagoon
-múli promontory
-sandur sand, sand plain
-skáli lodge
-skarð pass
-staður/staðir place
-strönd shore
-sýsla county
-vað ford
-varða cairn
-vatn lake
-vegur road, route
-vellir field
-öræfi barrens

Index

References

Ari Trausti Guðmundsson (2001): *Íslenskar eldstöðvar.*

Árbók F.Í. 1967: Sprengisandur.

Árbók F.Í. 1976: Fjallabaksleið syðri.

Árbók F.Í. 1981: Ódáðahraun.

Árbók F.Í. 1996: Ofan Hreppafjalla.

Árbók F.Í. 2000: Í strandbyggðum norðan lands og vestan.

Björn Hróarsson (1994–1995): *Á ferð um landið.*

Bragi Sigurjónsson (1948–1953): *Göngur og réttir 1-V.*

Friðrik Páll Jónsson (1983): *Út og suður. 20 ferðaþættir.*

Guðlaugur Jónsson (1983): *Bifreiðar á Íslandi 1904-1930.*

Guðmundur Kristinsson 2001): *Styrjaldarárin á Suðurlandi.*

Guðmundur Páll Ólafsson (2000): *Hálendið í náttúru Íslands.*

Ingibjörg Bergþórsdóttir (1994): *Söguferð bæ frá bæ um Borgarfjörð.*

Ingvar Teitsson (2005): *Biskupaleið yfir Ódáðahraun.*

Kristján Þórðarson (2006): *Vegir og vegleysur*

Magnús H. Helgason (ed.) (1995): *Saga Borgarfjarðar eystra.*

Ólafur Jónsson (1945): *Ódáðahraun.*

Ómar Ragnarsson (1999): *Ljósið yfir landinu.*

Páll Ásgeir Ásgeirsson (2001): *Hálendishandbókin.*

Páll Sigurðsson (1994) : *Fjallvegafélagið. Ágrip af sögu þess.*

Pálmi Hannesson og Jón Eyþórsson (1949–1957): *Hrakningar og heiðavegir 1-4.*

Steindór Steindórsson og Þorsteinn Jósepsson (1974): *Landið þitt.*

Tómas Einarsson og Helga Magnússon (ed.) (1989): *Íslandshandbókin.*

Útivera, a magazine.

Þorsteinn Matthíasson (1983): *Áfram skröltir hann þó. Páll Arason fjallabílstjóri rekur minningar sínar úr byggðum og óbyggðum.*

The Web:

http://steinisv.blog.is/blog/steinisv/image/399573

www.borgarfjordureystri.is

www.gamli.umhverfissvid.is/reykjanesfolkvangur/

www.hrauneyjar.is

www.kerlingafjoll.is

www.klaki.net/vegir

www.landmannalaugar.is

www.leif.is

www.nat.is

www.natturuvernd.is

www.sorli.is/greinar/gamlar_gotur_borgarf.htm
www.takgil.is
www.ust.is
www.vegur.is.
www.vik.is
www.visit.is